NANOTECHNOLOGY

MJP
PUBLISHERS

NANOTECHNOLOGY

Dr. S. Shanmugam

Consultant Diabetologist and Echocardiologist
and
Visiting Professor in Nanotechnology
Department of Advanced Zoology and Biotechnology
St. John's College
Palayamkottai, Tamilnadu

MJP Publishers

No. 44, Nallathambi Street,
Triplicane, Chennai 600 005

MJP 068 © Publishers, 2024

Publisher : C. Janarthanan

This book has been published in good faith that the work of the author is original. All efforts have been taken to make the material error-free. However, the author and publisher disclaim responsibility for any inadvertent errors.

To

The memory of the noble souls departed from our family in 2008

   S. Avudaiappan (Brother) & an innocent and childlike personality

   S. Kuppuswamy (Brother) & The best brain in the banking industry and a lovable brother

   T. Karthikeyan (Son-in-law) & A too good man with no enmity

The developments in Nanomedicine would have cured their diseases. Unfortunately we are only in the beginning of that era.

PREFACE

There is a quotes in Tamil—"*Thirukural*, the great book of Tamilians was created by drilling the mustard seed and infusing the seven seas of the world." This proves that the minuscule nature of manufacturing which is now called as Molecular Manufacturing was in the thinking level in Tamil Nadu thousands of years ago.

Nanotechnology can be defined as a field of applied science and technology which deals mainly with the control of matter on the atomic and molecular scale (normally 1 to 100 nm and the fabrication of devices within that size range. Applied physics, materials science, interface and colloid science, device physics, supramolecular chemistry and even mechanical and electrical engineering are the fields from which this highly multidisciplinary science has evolved for the past half a century. There is a great expectation and speculation regarding this new science and technology and these lines of research. Nanotechnology is considered as an extension of existing sciences into the nanoscale or putting it in a different way, it is recasting of existing sciences using a newer, more modern term.

In India, modest beginnings have been made to catch up with the nanotechnological innovations. Today, only a few institutions are contributing towards this pioneering research. A lot more needs to be done and an action plan is the necessity. Making India a significant player in Nanoscience and Nanotechnology should be the aim and goal of the intelligentsia of India. The introduction of Nanotechnology in the curriculum of various courses in India at colleges and universities recently made this science being known at the academic level. The study materials needed for the teachers to teach and the students to learn is really at the wanting level in the Indian context. If one sees the nanopublications scenario of the world, India has a long way to go.

This book aims to provide a basic understanding of the field for beginners whereas readers already exposed to this field may expand their knowledge in the subject. Being in the field of Nanomedicine research for the past two years, especially in relation to Homeomedicine as a Nanomedicine and also working on the utility of various metal vessels in Siddha Medicine as a form of Nanomedical application for cure of various diseases, prompted me to share my knowledge gained as a Nanomedicine researcher for the benefit of the teacher and student community of India, who are really in need of proper guidance in this growing field of science in the Indian context.

I took up to being a teacher in the emerging field of nanoscience, with a lot of joy and enthusiasm. To acquire knowledge effectively, synthesize the enormous data which are mostly hypothetical and theoretical in nature and communicate the information to the student community in such a way that they would understand, assimilate and reproduce easily in the examination is a great task. This tests one's ability as a teacher in the modern day scenario. That experience is the base for writing a book for students at all levels of competence—undergraduate, postgraduate, doctoral and postdoctoral levels and also for the teachers of Nanotechnology. Let us explore the Modern Molecular Science of today in the pages to come.

In India at the moment, there is a lack of a good and comprehensive textbook for the courses which require Nanotechnology to be taught in colleges and universities. This book may serve to fill this gap.

S. Shanmugam

CONTENTS

APPENDICES	**235**

Chapter 1

Introduction

Nanotechnology is a relatively recent development in scientific research. But the development of its central concepts happened over a longer period of time. Nanotechnology as the term used and defined today is not different from the basic usage for which the materials were used in the science long ago in different countries and in different applications. Only the theoretical concept is evolved recently in a classical science pattern. The evolution of the "nano" concept can be given in a chronological order (Table 1.1) and the evolution of modern nanotechnology is depicted in Figure 1.1.

Table 1.1   Evolution of nanoconcept

Discovery type	Name	Age	Year
Industrial	Tools	Stone	2,200,000 BC
Industrial	Metallurgy	Bronze	3500 BC
Industrial	Steam power	Industrial	1764
Automation	Mass production	Consumer	1906
Automation	Computing	Information	1946
Health	Genetic Engineering	Genetic	1953
Industrial	Nanotechnology	Nano Age	1991
Automation	Molecular assemblers	Assembler Age?	2020?
Health, industrial, automation	Life assemblers	Life Age	2050?

Source: Wilson et al. 2002. *Nanotechnology: Basic Science and Emerging Technologies.* Chapman and Hall/CRC, New York.

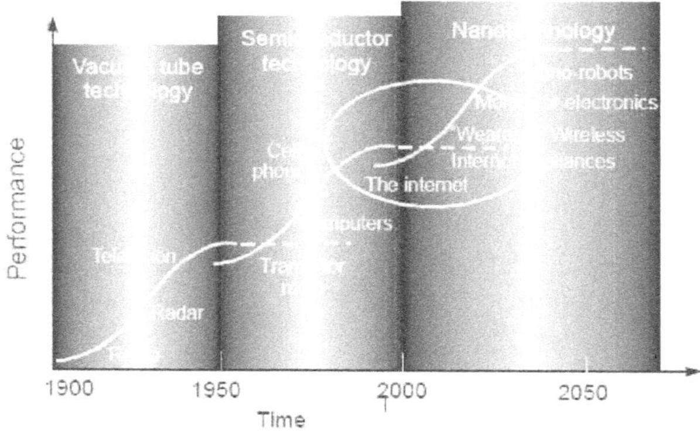

Figure 1.1   Evolution of modern nanotechnology
(*Source:* USDA, Dr. Hongda Chen's presentation)

PRE-NANOTECHNOLOGY

The employment of nanotechnology in human activities dates back to thousands of years. Some examples are the making of steel and vulcanizing rubber. Both these processes depend on the properties of **stochastically**[1] formed atomic ensembles with size in nanometres. They are distinguished from chemical reaction since they are not depending on the properties of individual molecules. In spite of the use of nanotechnology for so many centuries, the development of the concepts which are considered under the term nanotechnology has been slower.

James Clerk Maxwell

In 1867, James Clerk Maxwell proposed a hypothetical concept, a tiny entity known as **Maxwell's Demon**. It has got the capacity to handle individual molecules. This was the first and the beginning of some of the distinguishing concepts in nanotechnology. But this pre-dated the use of the name nanotechnology.

During the first decade of the 20th century, observations and size measurements of nanoparticles were made for the first time. The credit for such work is mostly associated with the name Zsigmondy. He made a detailed study of gold sols and other nanomaterials with sizes down to 10 nm and less and published a book in 1914. The instrument used by him was the ultramicroscope that employs dark-field method for seeing particles with sizes much less than the wavelength of light.

Richard Adolf Zsigmondy

Zsigmondy is also credited with coining the term nanometre explicitly for characterizing particle size. He was the one who determined it as 1/1,000,000 of a millimetre. He was also the first to develop a system of classification based on particle size in the nanometre range.

Many significant developments have been made during the 20th century in characterizing the nanomaterials and related phenomena. Interface and colloid science is the scientific discipline related to nanotechnology.

The concept of a monolayer, a layer of material one molecule thick, was introduced in the 1920s by Irving Langmuir and Katharine B. Blodgett, and Langmuir won the Nobel Prize in Chemistry for this work.

The first measurement of surface forces was conducted by Derjaguin and Abrikosova in the early 1950s. Studies of periodic colloidal structures and principles of molecular self-assembly have been many.

Fundamentals of Interface and Colloid Science by H. Lyklema is the source book for discoveries that serve as the scientific basis for modern nanotechnology.

ORIGINS OF CONCEPTS OF NANO

Nanoscience and nanotechnology, owe a lot to two personalities. First is physicist Richard Feynman. He described the concept of "building machines" atom by atom in his talk titled "There is plenty of room at the bottom" at Caltech on December 29, 1959. In that lecture, Feynman described a process, which is now known as nanotechnology as follows: "the process by which the ability to manipulate individual atoms and molecules might be developed, using one set of precise tools to build and operate another proportionally smaller set, so on down to the needed scale." This description is the basis and the first use of the distinguishing concepts in "nanotechnology" before the term was coined. Scaling issues that would arise from the changing magnitude of various physical phenomena were also noted by him. Thus, gravity would become less important, surface tension and **van der Waals attraction**[2] would become more important and the other properties also change. This basic idea was accepted as feasible. Exponential assembly enhances the capacity of nanotechnology's potential with parallelism. This in turn can produce a useful quality of end products. This capacity enhancement possibility was discussed.

The second one is Dr. K. Eric Drexler[3], who by writing the book titled *Nanosystems: Molecular Machinery, Manufacturing and Computation*, created the basis of molecular manufacturing which is an important component of nanoscience.

The construction of a nanomotor was achieved by William McLellan in November 1960. Scaling down the letters small enough so as to be able to fit the entire *Encyclopedia Britannica* on the head of a pin was achieved in 1985 by Tom Newman.

In 1965, Gordon Moore created Moore's law which was due to the observation that silicon transistors were undergoing a continual process of scaling downward. Moore's law was responsible for the decrease of transistor "minimum feature" sizes from 10 μm to 45–65 nm range in 2007. (One minimum feature is roughly 180 silicon atoms long).

Professor Norio Taniguchi of Tokyo Science University was the first to describe the term "nanotechnology" in 1974.[4] In his words, "Nanotechnology mainly consists of the process of separation, consolidation and deformation of materials by one atom or by one molecule." From then onwards the definition of nanotechnology has generally been extended upward in size to include features as large as 100 nm. Apart from that, it also embraces structures exhibiting quantum mechanical aspects, such as quantum dots and has been included into the definition.

Dr. Tuomo Suntola and co-workers in Finland developed and patented the process of atomic layer deposition, for depositing uniform thin films of one atomic layer at a time in 1974.

Dr. K. Eric Drexler promoted the technological significance of nanoscale phenomena and devices through speeches. In the 1980s, the basic idea of nanotechnology as deterministic, rather than stochastic, i.e., handling of individual atoms and molecules was conceptually explored in depth. *Engines of Creation: The Coming Era of Nanotechnology* (1986) and *Nanosystems: Molecular Machinery, Manufacturing, and Computation* (1998) are the two books written by Dr. K. Eric Drexler in which he has described the concepts well. Because of this effort, the term acquired its current sense. Drexler's vision of nanotechnology is equated with "molecular nanotechnology" (MNT) or "molecular manufacturing." Drexler proposed the term "zettatech" which never became popular.

ADVANCES IN EXPERIMENTAL METHODS

In the 1980s, nanotechnology and nanoscience became more popular with two major developments.

1. The recognition of cluster science as a field and the invention of the scanning tunnelling microscope (STM). This development led to the discovery of fullerenes[5] in 1985 and the structural assignment of carbon nanotubes a few years later.

2. In the second major development, the synthesis and properties of semiconductor nanocrystals were studied. Number of semiconductor (metal oxide) nanoparticles

of quantum dots were synthesized in a faster way. The atomic force microscope was invented five years after the invention of STM.

In the early 1990s, Huffman and Kraetschmer (University of Arizona) discovered the means of synthesizing and purifying large quantities of fullerenes. This lead to their characterization and functionalization of fullerenes by a large number of investigators. Within a short span, rubidium-doped C_{60} was found to be a mid-temperature superconductor. In 1992, Dr. T. Ebbesen discovered and characterized the carbon nanotubes. Further development of the field of nanotube-based nanotechnology became feasible.

At present, the practice of nanotechnology is having both stochastic approaches (supramolecular chemistry creates waterproof pants) and deterministic approaches[6] (complex, deterministic molecular nanotechnology is at the moment attainable practically).

From the 1990s, surface scientists and thin film technocrats have redefined their disciplines as nanotechnology. This has caused a great deal of confusion as to what nanotechnology really means, and thousands of research articles relating to this have been published. Most of such research are extensions of the ordinary research done in the parent fields.

NANOTECHNOLOGY—BASICS AND BASIS

Nanotechnology is the engineering of functional systems at the molecular scale. Nanotechnology is an area of modern science and is defined as the science which deals with the ability to control and manipulate matter at a scale ranging from less than a nanometre up to 100 nm. One nanometre is one-billionth of a metre or 10^{-9} of a metre, roughly the width of three or four atoms. For comparison, typical carbon–carbon bond lengths or the spacing between these atoms in a molecule, are in the range 12–15 nm, and a DNA double-helix has a diameter around 2 nm. But the smallest cellular life forms, the bacteria of the genus *Mycoplasma*, are around 200 nm in length. To be precise, the comparative size of a nanometre to a metre is the same as that of a marble to the size of the earth. A nanometre may be considered as the amount a man's beard grows during the time it takes him to raise the razor to his face. Nano is a word derived from Greek, and it means one billionth part of a whole. A single human hair is around 100,000 nm in width, which will explain and make one to understand about the nanoscale parameter. In processes of nanotechnology, the main criterion is the manipulation of matter at or near the atomic and molecular level of precision.

Simply, nanotechnology is defined as the science which deals at the molecular level to create new materials by using functional systems which are engineered. This definition covers both present work and concepts that are more advanced and of future applications.

In its original meaning nanotechnology refers to the projected ability to construct items *from the bottom up*, using techniques and tools being developed today to make comple te, high performance products.

Size of Nano

$$1 \text{ nm} = 10^{-3}\,\mu\text{m} = 10^{-6}\,\text{mm} = 10^{-9}\,\text{m}$$

Some examples of sizes from the macro to the molecular level are provide in Table 1.2.

Table 1.2 Size and name of particles

Size (nm)	Examples	Terminology
0.1–0.5	Individual chemical bonds	Molecular/atomic
0.5–1.0	Small molecules, pores in zeolites	Molecular
1–1000	Proteins, DNA, inorganic nanoparticles	Nano
10^3–10^4	Microfluidic channels, MEMS, devices on a silicon chip, living cells (bacteria: 1 µm; yeast: 5 µm; human hair: 50 µm)	Micro
$>10^4$	Normal bulk matter	Macro

Source: USDA's roadmap of nanotechnology.

The Meaning of Nanotechnology

In the 1980s, K. Eric Drexler described nanotechnology as the building of machines of few nanometres width like motors, robot arms and whole computers, on the scale of molecules, far smaller than a cell. Drexler worked on these incredible devices which were hypothesized by him for the following ten years. He was described as a science fiction writer and not accepted as a scientist. But the ability to build simple structures on a molecular scale has become a reality and now nanotechnology has become an accepted concept. Nanotechnology is used to encompass the simpler kinds of nanometre-scale technology.

Whatever present work is being carried out under the name "nanotechnology" does not much pertain to the original meaning of the word. Nanotechnology is building things from the bottom up, with atomic precision in its traditional sense. As early as 1959, the renowned physicist Richard Feynman envisioned this theoretical capability and possibility.

"I want to build a billion tiny factories, models of each other, which are manufacturing simultaneously. The principles of physics, as far as I can see, do not speak against the possibility of maneuvering things atom by atom. It is not an attempt to violate any laws; it is something, in principle, that can be done; but in practice, it has not been done because we are too big."

Richard Feynman

Advanced nanotechnology or molecular manufacturing will make use of positionally controlled mechanochemistry guided by molecular machine systems. This is on the basis of Feynman's vision of miniature factories using nanomachines to build complex products.

Four Generations of Nanotechnology Development

Four generations of nanotechnology development have been described by Mihail (Mike) Roco of the U.S. National Nanotechnology Initiative.

1. **Current era**   It is the first phase in which the development of passive nanostructures, materials designed to perform one task, is happening. This is otherwise called as **basic era.**

2. **Development era**   The second phase deals with active nanostructures for multitasking. For example, actuators, drug delivery devices, and sensors.

3. **Future era**   The third generation or phase is expected to emerge by 2010. It will feature nanosystems with thousands of interacting components.

4. **System era**   The phase where there is development of a mechanical system like that of biological system. The first integrated nanosystems functioning much like a mammalian cell with hierarchical systems within systems is the goal of this era.

There are views that nanotechnology can refer to measurement or visualization at the scale of 1–100 nm only. But control and restructuring of matter at the nanoscale as a necessary element is the view that is now accepted. As stages of four generations of nanotechnology progress, it will lead to molecular nanosystems. This also includes molecular manufacturing. Ultimately "engineering of functional systems at the molecular scale" is what nanotech is expected to deliver for applications.

Technology of General Applicability

Nanotechnology can be considered as a technology of general applicability (General-purpose technology), since in its advanced era, its significant impact will be on almost all industries and all areas of society. It is expected to offer better built, longer lasting, cleaner, safer and smarter products for the home, for communication, for medicine, for transportation, for agriculture and for industry in general.

Quoting from the reports of the U.S. National Science Foundation will be of use in understanding the above definition.

Imagine a medical device that travels through the human body to seek out and destroy small clusters of cancerous cells before they can spread. Or a box no larger than a sugar cube that contains the entire contents of the Library of Congress. Or materials much lighter than steel that possess ten times as much strength.

U.S. National Science Foundation

Multi-purpose Technology

As the previous years' improvement in human life was due to electricity or computers as example, in future, nanotechnology is going to offer greatly improved efficiency in almost every walk of human life. It has a multi-purpose role. That means it will be utilized in many commercial uses and also many military uses—making far more powerful weapons and tools of surveillance. Thus it is a double-edged technology representing not only wonderful benefits for humanity, but also grave risks.

A key advantage of nanotechnology is that it offers not just better products, but a vastly improved manufacturing process. For example, a computer can make copies of data files—essentially as many copies as needed at little or no cost. It is only a matter of time by which, the building of products becomes as cheap as the copying of files. That explains the real meaning of nanotechnology, and it is predicted as "the next industrial revolution. "The nanotechnology revolution has the potential to change the world on a scale equal to, if not greater than, the computer revolution" is the apt comment about nano.

A personal nanofactory resting on countertop in a shop or desktop in an office or house, in which the power of nanotechnology is being incorporated, is possible with the present vision regarding nanotechnology. Such personal nanofactory which is packed with miniature chemical processors, computing and robotics can produce a wide range of items quickly, cleanly, and inexpensively. Building products directly from blueprints on the place of work will be the order of this future technology.

Exponential Proliferation

Nanotechnology's possibilities are that it will not only allow making many high-quality products at very low cost but also will allow making new nanofactories at the same low cost and at the same rapid speed. This unique ability to reproduce its own means of production, which is seen in nature in biological processes only, is the reason for calling nanotech as an exponential technology. It is a manufacturing system that makes more manufacturing systems or is a self-replicating system—factories that can build factories rapidly, cheaply and cleanly. The means of production has been predicted to reproduce exponentially. That means, in just a few weeks, a few nanofactories conceivably could become billions. It is considered as a revolutionary, transformative, powerful and potentially very dangerous—or beneficial—technology.

What is the time line for all these to happen? Predicted at present, it is 20 to 30 years from now or even much later than expected. The development can be sooner than predicted quite possibly within the next decade. This may be due to the rapid progress being made in enabling technologies, such as optics, nanolithography, mechanochemistry and 3D prototyping. If nanotechnological application could come so soon, the consequences due to that quick arrival could be severe or disastrous to the human race, since the world in not adequately prepared.

INSIDE THE NANO WORLD

Molecular Manipulation for Future Unbelievable Materials

Materials technology adds to the real wealth of nations. Studying the civilizations stretching back 8,000 years to the Bronze Age, it is the materials which form the hallmark of studying the civilizations. The present Information Age is not different regarding the influence of materials. The basis of computer chips, cell phones or fibre-optic networks is nothing but silicon, that is 99.99999% pure. In the past decades, inorganic chemists have created an impressive array of metals, alloys, and ceramics. Those are the materials responsible for sending skyscrapers ever higher, making cars lighter and more fuel-efficient and launching air travel for the masses.

Organic chemistry contributed to modern conveniences, on its part. It is responsible for the production of the plastic packages that keep foods fresh, the housings that protect TVs and appliances and synthetic woolen materials that ward off the cold. Organic chemists created the fertilizers, that helped in the Green Revolution and unlimited and unimaginable wonder drugs that make people healthy.

Presently materials scientists are researching to transform the world once again. From the utilization of raw materials extracted from the earth, researchers are squeezing their brains for totally new and ultramodern structures. This is possible by tearing down the walls between organic and inorganic chemistry. These happenings would have been dismissed as nonentities long ago.

Biological Molecular Complexity

From giant stars to human bodies in this universe, everything works on a molecular scale. Nanophysiology, a branch of nanotechnology, is dealing with all the processes taking place at the molecular level in the heart, lungs and other organs of the human body in spite of they being visually big objects. Everything in the human body as well as in the physical universe is already taking place based on nanotechnology. Observation of the nature and the role of science in understanding it from the research in nanosciences can be converted into a technological product. This can be done by using the same or similar nanomaterials which gave the natural colour for the peacock feather, as part of the shirts, sarees, fabric and apparels which human beings wear. It is a welcome and future destination for science to mature into technology and become a product of utility for the society. Figure 1.2 is an illustration of te vision of nanoscale science and technology as a botom-up paradigm.

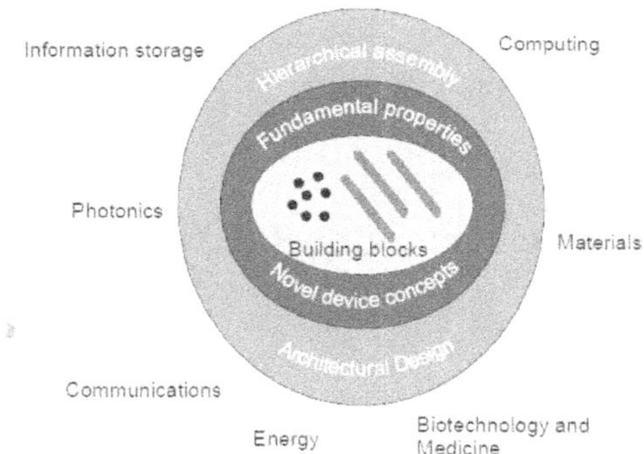

Figure 1.2   Vision of nanoscale science and technology as a bottom-up paradigm

Dimensions of Nanoscience and Technology

In the last 150 years, humanity has seen rapid transformation in the manufacturing sector. This revolution has continued with the introduction of numerically controlled robotics. Of late, in the manufacturing of materials, new trends with the introduction of software and design tools are witnessed. The phenomenal progress in microelectronics during the last six decades has added a new dimension to manufacturing. Considerable progress from microprocessor in the 1970s to MEMS[7] in 2000 and now to nanotechnology is made. Nanofabrication, biocomputing, molecular computing and quantum computing are the future manufacturing methodologies.

The application of nanotechnology to the day-to-day living scenario is growing in the world. That makes it mandatory to understand the necessity to master the newer science at the earliest for any country to develop in technology and reach the financial goals. The world market for nanomaterials, nanotools, nanodevices and nanobiotechnology put together is expected to be in billion dollars and is growing day by day. The fastest growing area among these is nanobiotechnology and is to be noted for planning the future technology.

The emerging areas of Nanotechnology and their applications need some introduction before going into the details in the pages to come.

1. **CNT and composites**   Super strong, smart and intelligent structures in the field of material science are expected to be developed out of carbon nanotubes and its composites, making a revolution in the materials science arena.

2. **Nanocomputers**   The development of molecular switches and circuits along with nanocell will give rise to the next generation computers. Ultradense computer memory coupled with excellent electrical performance will lead to low-power, low-cost, nanosized and yet faster assemblies, which will be useful to the society.

3. **Bio-nanotechnology**   Nanobiomedical sensors will play a major role in glucose detection especially in diabetics and endoscopic implants in various organs of the body. The newer drug delivery system under development will revolutionize health care in a big way in future.

4. **Packaging**   The last four decades changed the packaging concept totally. From electronics packaging of the past to the present microsystems packaging, trend is towards futuristic nanopackaging.

5. **Nanorobots**   The field of mechano-compatibility for the human body is an area of interest for doctors and medical experts in the past few years. The idea is to create a nanorobot within the human body which can be operated both chemically and physically. Currently biomedical instruments and procedures to explore tissues and cells are available for the physicians. A new field of exploring the human body with nanorobots is of great interest to doctors and medical scientists. Thus these nanorobots to be developed, must pass through a person's body without side effects like excessive bruising, itching and other disturbances. Finding a way for the nanorobots to perform the maximum amount of biomedical operations with the least amount of irritation and other possible illnesses to the patient is the key for success in this fast emerging field of study.

6. **Soft lithography and nanoscale printing**   Soft lithography is generally used to construct features measured on the nanometre scale. It has some unique

advantages over other forms of lithography like that of photolithography and electron beam lithography.

7. **Fuel cell** Nanotechnology-based fuel cells hold a promising future for fuel automotive industry, and for power grids and batteries for all types of applications—mobile, portable and stationary use.

Globally, fast advances are taking place both in terms of materials and devices. Around 300 novel materials and 50 devices with unique characteristics have been successfully developed around the world within a short span of time. Many of these inventions in materials and devices are being created today by nanotechnology and which were beyond human imagination and comprehension few years back. Globally, large number of universities, academic institutions and companies are making serious and concrete efforts in the direction of nanotechnology and the applicability of nanomaterial.

Nano is the greatest building block for health care, structural material, in electronics, automation, etc. This new science will become the platform for launching new cutting-edge technologies for the better living of mankind.

DEFINITIONS IN NANOTECHNOLOGY

The varied and conflicting definitions given to nanotechnology and the unclear boundaries between the different fields of science make it difficult for one to have a clear understanding of nanoscience and of the differences between this field and the other fields of science. This also hinders the effective and sensible development policy regarding nanotechnology.

Today, nanotechnology is concerned with a broad collection of mostly disconnected fields. Literally, anything sufficiently small and interesting can be considered under nanotechnology. Much of what is called nano today is harmless. But with the available data, much of the harm is of familiar and of limited quality. But surprising, unfamiliar and newer risks and problems are expected out of molecular manufacturing.

The various terms related to nanotechnology are given.

Nanobiotechnology Bio-nanotechnology is concerned with molecular scale properties and applications of biological nanostructures and as such it sits at the interface between the chemical, biological and the physical sciences like a cat on the wall.

Nanocomposites Nanomaterials that combine one or more separate components in order to obtain the best properties of each component (composite). In nanocomposite, nanoparticles (clay, metal, carbon nanotubes) act as fillers in a matrix, usually polymer matrix.

Nanoelectronics Application of nanoscience and nanotechnology techniques in the field of electronics are especially promising in computer chips, optoelectronics, information storage, nanocomputer, and in sensors.

Nanomachines Devices which are built from individual atoms and their size is measured in nanometres. The idea is that the assembler will be able to rearrange the atoms from raw material in order to produce useful items.

Nanomanufacturing Manufacturing at the nanoscale. It is the industrial application of nanotechnology.

Nanomaterials Materials which have structured components with at least one dimension less than 100 nm. Materials that have one dimension in the nanoscale are layers, such as thin films or surface coatings. Some of the features on computer chips come in this category. Materials that are nanoscale in two dimensions include nanowires and nanotubes. Materials that are nanoscale in three dimensions are particles, for example, precipitates, colloids and quantum dots (tiny particles of semiconductor materials). Nanocrystalline materials, made up of nanometre-sized grains, also fall into this category.

Nanomedicine Application of nanoscience and nanotechnology techniques in the field of medicine. Areas such as disease diagnosis, drug delivery and molecular imaging are being intensively researched. Medical-related products containing nanoparticles are currently produced.

Nanometre One nanometre (nm) is equal to one-billionth (10^{-9}) of a metre. Atoms are below a nanometre in size, whereas many molecules, including some proteins, range from a nanometre upwards.

Nanometrology The science of measurement at the nanoscale level. Nanometrology has a crucial role in the production of nanomaterials and devices with a high degree of accuracy and reliability (nanomanufacturing).

Nanoparticle Particles of less than 100 nm in diameter that exhibit new or enhanced size-dependent properties compared with larger particles of the same material.

Nanoscale Scale with nanometre order of magnitude.

Nanoscience The study of the phenomena and manipulation of materials at the atomic, molecular and macromolecular scales, where the properties significantly differ from those at a larger scale.

Nanostructure A structure with arrangement of its parts in the nanometre scale.

Nanotechnologies The design, characterization, production and application of structures, devices and systems by controlling shape and size at the nanometre scale.

Nanotube (Carbon nanotubes)   Carbon nanotubes (CNTs) were discovered by Sumio Iijima in 1991. Carbon nanotubes are fullerene-related structures which consist of rolled graphene sheets. There are two types of CNT: single-walled (one tube) and multi-walled (more tubes). Both of these are typically a few nanometres in diameter and several micrometres to centimetres long.

APPLICATIONS OF NANOTECHNOLOGY

The list of applications of nanotechnology is growing day by day due to the ongoing research all over the world. In few years or a decade time, today's theoretical possibilities in nanotechnology will become practical reality. Whatever is considered as fiction today will be future achievements in the field of nanotechnology.

With nanotechnology, materials and products will have changes in the physical properties due to shrinking of size. Nanoparticles with their increased surface-area-to-volume ratio are advantageous, an example being fluorescence changing as per the diameter of the nanoparticles. When associated with a bulk material, nanoparticles influence the mechanical properties of the material, like stiffness or elasticity. As an example, traditional polymers can be reinforced by nanoparticles resulting in novel light-weight materials which would be alternatives to metals, presently used. Nanotechnologically enhanced materials will enable a weight reduction, increased stability and improved functionality.

There are plenty of fields in which the applications of nano are well known during the past years and such applications are growing day by day. The applications of nanotechnology in various fields of study are discussed.

Nanomedicine

Biological and medical research utilized the unique properties of nanomaterials for various applications like contrast agents for cell imaging and therapeutics for treating cancer. Biomedical nanotechnology, bionanotechnology, and nanomedicine are the other nomenclatures for this hybrid field. Functionalities can be added to nanomaterials by interfacing them with biological molecules or structures. The size of nanomaterials is similar to that of most biological molecules and structures. Because of this property, nanomaterials can be useful for both *in vivo* and *in vitro* biomedical research and applications. The integration of nanomaterials with biology has led to the development of diagnostic devices, contrast agents, analytical tools, physical therapy applications and drug delivery vehicles.

Diagnostics   Nanotechnology-on-a-chip is a form of lab-on-a-chip technology. Biological tests—measuring the presence or activity of selected substances—become quicker, more sensitive and more flexible when certain nanoscale particles are put to work as tags or labels. Some specific applications of nanotechnology in diagnostics are:

1. Magnetic nanoparticles, bound to a suitable antibody can be used to label specific molecules, structures or microorganisms.

2. Gold nanoparticles tagged with short segments of DNA can be used for detection of genetic sequence in a sample.

3. Multicolour optical coding for biological assays has been achieved by embedding different-sized quantum dots into polymeric microbeads.

4. Nanopore technology for analysis of nucleic acids converts strings of nucleotides directly into electronic signatures.

Drug delivery systems   The drug dosage as well as side-effects of the drugs can be reduced by depositing the active drug in the diseased tissues only, thereby reducing the need of higher doses. This target-oriented approach saves time and money and reduces human suffering.

Nanotechnology-oriented implantable drug delivery system is preferable to the use of injectable drugs. Injectables work on first-order kinetics namely the blood concentration goes up rapidly, but drops exponentially over time. This will increase drug toxicity, and drug efficacy is unpredictable as the drug concentration falls below the targeted range.

Some specific applications in drug delivery systems are:

1. Dendrimers and nanoporous materials could hold small drug molecules transporting them to the desired location.

2. Applications based on small electromechanical systems such as NEMS are researched for the active release of drugs. Important applications of this include treatment for cancer with iron nanoparticles or gold shells.

Tissue engineering   Nanotechnology can help to reproduce or to repair damaged tissue. This "tissue engineering" makes use of artificially stimulated cell proliferation using suitable nanomaterial-based scaffolds and growth factors. Tissue engineering might replace conventional treatments like organ transplants or artificial implants. Advanced nanotechnology-based tissue engineering can also lead to longevity of life in a matter of years. But, the concept of tissue engineering is a matter of debate on ethical grounds like that of human stem cells.

Chemistry and Environment

Chemical catalysis and filtration techniques are the areas in chemistry in which nanotechnology plays an important role. The synthesis provides novel materials with tailored features and chemical properties like nanoparticles with a distinct chemical surrounding (ligands), or specific optical properties. In this aspect, chemistry is to be considered as a basic nanoscience. Chemistry will provide novel "nanomaterials" in short term and superior processes such as "self-assembly" will enable energy and time-preserving strategies in the long run. All chemical syntheses can be understood in terms of nanotechnology, because of its ability to manufacture certain molecules, and chemistry forms a base for nanotechnology providing tailor-made molecules, polymers as well as clusters and nanoparticles.

Catalysis Catalysis is important for the production of chemicals. Chemical catalysis benefits from nanoparticles, due to the large surface-area-to-volume ratio. The application of nanoparticles in catalysis is from fuel cell to catalytic converters and photocatalytic devices.

Filtration Influence of nanochemistry on waste-water treatment, air purification and energy storage devices is on the anvil. Mechanical as well as chemical methods can be used for effective filtration techniques. Nanoporous membranes are suitable for a mechanical filtration with extremely small pores smaller than 10 nm (nanofiltration) and may be composed of nanotubes. Nanofiltration is mainly used for the removal of ions or the separation of different fluids. The membrane filtration technique is named ultrafiltration, which works down to between particles of size 10 and 100 nm. One important field of application of ultrafiltration for medical purposes is in renal dialysis. By making use of magnetic separation techniques, magnetic nanoparticles offer an effective and reliable method to remove heavy metal contaminants from waste water. Nanoscale particles increase the efficiency to absorb the contaminants and this method is inexpensive compared to traditional precipitation and filtration methods.

Energy

The most advanced nanotechnology projects related to energy are 1) storage, 2) conversion, 3) improving manufacturing process by reducing materials and process rates, 4) energy saving by better thermal insulation, and 5) enhanced renewable energy sources.

Reduction of energy consumption A reduction of energy consumption is possible by better insulation systems, by the use of more efficient lighting or combustion systems and by use of lighter and stronger materials in the transportation sector. Light bulbs presently used convert only approximately 5% of the electrical energy into light. Nanotechnological approaches like light-emitting diodes (LEDs) or quantum-caged atoms (QCA) could lead to a low energy consumption for illumination.

Increasing the efficiency of energy production Best solar cells manage to use 40 per cent of the sun's energy in spite of layers of several different semiconductors stacked together to absorb light at different energies. Commercially available solar cells have much lower efficiencies (15–20%). Tetrad-shaped nanoparticles when applied to a surface instantly transform it into a solar collector. Nanotechnology could help increase the efficiency of light conversion by using nanostructures with a continuum of band gaps.

The degree of efficiency of the internal combustion engine is about 30–40%. Nanotechnology could improve combustion by designing specific catalysts with maximized surface area.

More environment-friendly energy systems Environment-friendly form of energy is the use of fuel cells powered by hydrogen, which is ideally produced by renewable energies. Important nanostructured material in fuel cells is the catalyst consisting of carbon-supported noble metal particles with diameters of 1–5 nm. Suitable materials for hydrogen storage contain a large number of small nanosized pores. Many nanostructured materials like nanotubes, zeolites or alanates are under research. Nanotechnology can contribute to the further reduction of combustion engine pollutants by nanoporous filters which can clean the exhaust mechanically, by catalytic converters based on nanoscale noble metal particles or by catalytic coatings on cylinder walls and catalytic nanoparticles as additive for fuels.

Recycling of batteries The disposal of huge quantities of used batteries and accumulators is a great problem. The use of batteries with higher energy content or the use of rechargeable batteries or super capacitors with higher rate of recharging using nanomaterials could be helpful for the safe disposal of these batteries.

Information and Communication

In high-technology production processes, the introduction of nanotechnology has already occurred. The gate length of transistors used in CPUs or DRAM is at the nanoscale (50 nm or below) especially the critical length scale of integrated circuits used in them.

Memory storage Electronic memory designs in the past have largely relied on the use of transistors. Research of crossbar switch-based electronics has provided an alternative using reconfigurable interconnections between vertical and horizontal wiring arrays to create ultra-high density memories. Two types of memory storage are:

1. a carbon-nanotube-based crossbar memory called nano-RAM.
2. the use of memristor material as a future replacement of flash memory.

Novel semiconductor devices Novel semiconductor devices are based on spintronics. The dependence of the resistance of a material (due to the spin of the electrons) on an external field is called magnetoresistance. This effect can be significantly amplified for nanosized objects. This giant magnetoresistance (GMR) effect has led to a strong increase in the data storage density of hard disks up to gigabytes. The tunnelling magnetoresistance (TMR) is similar to GMR and based on the spin-dependent tunnelling of electrons through adjacent ferromagnetic layers. Both GMR and TMR effects can be used to create a non-volatile main memory for computers, such as magnetic random access memory (MRAM).

Novel optoelectronic devices   In modern communication technology, traditional analog electrical devices are increasingly replaced by optical or optoelectronic devices due to their enormous bandwidth and capacity, respectively. Two areas of importance are photonic crystals and quantum dots.

1.   Photonic crystals are materials with a periodic variation in the refractive index with a lattice constant that is half the wavelength of the light used. They offer a selectable band gap for the propagation of a certain wavelength and they resemble a semiconductor, but for light or photons instead of electrons.

2.   Quantum dots are nanoscaled objects used for the construction of lasers. The advantage of a quantum dot laser over the traditional semiconductor laser is that their emitted wavelength depends on the diameter of the dot. Quantum dot lasers are cheaper and offer a higher beam quality than conventional laser diodes.

Displays   The production of displays with low energy consumption is possible using carbon nanotubes (CNT). Carbon nanotubes can be electrically conductive and can be used as field emitters with extreme high efficiency for field emission displays (FED) due to their small diameter of several nanometres. The principle of operation is like the cathode ray tube on a much smaller length scale.

Nanologic   Nanoscale devices exhibit dominant nonlinearities that prevent their use as two-state devices in digital computers. The idea behind nanologic is to exploit these nonlinearities instead of suppressing them to implement functions that correspond to mathematical sets like interval numbers, disjoint intervals, fuzzy numbers, fuzzy sets, etc. Simple nanoelectronic circuits can be designed that can represent sets and set operations, and an array of such devices constitutes a universal mathematical processor, able to solve any problem that can be expressed in set theory. Nanologic will help to solve problems involving uncertainty, ambiguity, error, under-specified and over-specified systems, and for approximate analysis of combinatorially intractable problems, including mathematical theorems, etc.

Quantum computers   New approaches for computing use the laws of quantum mechanics for novel quantum computers, which enable the use of fast quantum algorithms. The quantum computer will have quantum bit memory space termed qubit for performing several computations at the same time.

Heavy Industry

An inevitable use of nanotechnology will be in the heavy industry.

Aerospace   Lighter and stronger materials will be of immense use to aircraft manufacturers, leading to increased performance. Spacecraft, where weight is a major factor, will also benefit. Nanotechnology thereby helps to reduce the size of equipment used.

Through the use of nanotech materials, the weight of hang-gliders are reduced considerably while increasing their strength and toughness. Nanotechnology helps in lowering the mass of supercapacitors that are used to give power to assistive electrical motors for launching of hang-gliders.

Construction   Nanotechnology can make construction faster, cheaper, safer, and more varied. Automation of nanotechnology construction can allow for the creation of structures from advanced homes to massive skyscrapers much more quickly and at much lower cost.

Refineries   Refineries that produce materials such as steel and aluminium will be able to remove any impurities in the materials with the application of nanotechnology.

Vehicle manufacturers   Lighter and stronger materials will be useful for creating vehicles which are faster and safer. Combustion engines will also benefit from parts that are more wear- and heat-resistant.

Consumer Goods

Nanotechnology is making an impact in the field of consumer goods, providing products with novel functions ranging from easy-to-clean to scratch-resistant.

Foods   Nanotechnology can be utilized in the production, processing, safety and packaging of food. A nanocomposite coating process could improve food packaging by placing anti-microbial agents directly on the surface of the coated film. Nanocomposites could increase or decrease gas permeability of different fillers as is needed for different products. They can also improve the mechanical and heat-resistance properties and lower the oxygen transmission rate. Research is being performed to apply nanotechnology to the detection of chemical and biological substances for sensing biochemical changes in foods.

Household   The nanotechnology applications in the household items and equipment aims at self-cleaning or "easy-to-clean" surfaces on ceramics or glasses. Common household equipment like flat irons when made with nanoceramic particles have improved the smoothness and heat resistance.

Optics   The sunglasses using protective and antireflective ultrathin polymer coatings are available. For optics, scratch-resistant surface coatings based on nanocomposites are used. Nano-optics could allow for an increase in precision of pupil repair and other types of laser eye surgery.

Textiles   The use of engineered nanofibres makes clothes water- and stain-repellent or wrinkle-free. Textiles with a nanotechnological finish can be washed less frequently and at lower temperatures. Nanotechnology has been used to integrate membrane containing tiny carbon particles and guarantee full-surface protection from electrostatic charges for the wearer. Modern textile products are wrinkle-resistant and stain-repellent. Soon, clothes will become "smart", through embedded "wearable electronics".

Cosmetics   Especially in the field of cosmetics, nanoparticle-improved products have a promising potential. The traditional chemical UV protection approach suffers from its poor long-term stability. A sunscreen based on mineral nanoparticles such as titanium dioxide offer several advantages. Titanium oxide nanoparticles have a comparable UV protection property as the bulk material, but lose the cosmetically undesirable whitening as the particle size is decreased.

CONCLUSION

Nanotechnology is the engineering of tiny machines. It is the projected ability to build things from the bottom up, inside personal nanofactories (PNs), using techniques and tools being developed today to make complete, highly advanced products. The ultimate goal of nanotechnology is to control the matter at the nanometre scale, using mechanochemistry. Once this envisioned molecular machinery is created, it will result in a manufacturing revolution, probably causing severe social, physical, financial and so many other disruptions in human life. It also has serious economic, social, environmental, and military implications.

For the future, molecular nanotechnology (MNT) design evolution at the nanoscale which mimics the process of biological evolution at the molecular scale is to be developed. Biological evolution is by random variation in organisms combined with removal of the less-successful and reproduction of the more-successful variants. It is otherwise called application of Darwin's theory of evolution. Macroscale engineering design also proceeds by a process of design evolution from simplicity to complexity. The difficulty of seeing and manipulating at the nanoscale compared to the macroscale makes deterministic selection of successful trials difficult. In contrast biological evolution prceeds via action comprising random molecular variation and deterministic reproduction/extinction, called as "blind watchmaker."

ENDNOTES

1. A stochastic process is one whose behaviour is non-deterministic in that a state does not fully determine its next state. Stochastic crafts are complex systems

whose practitioners, even if complete experts, cannot guarantee success. Classical examples of this are medicine: a doctor can administer the same treatment to multiple patients suffering from the same symptoms, however, the patients may not all react to the treatment the same way. This makes medicine a stochastic process. Additional examples are warfare and rhetoric, where the successes and failures cannot be certainly predicted.

2. van der Waals attraction   Electrodynamic forces arise between atoms, molecules and assemblies of molecules due to their vibrations giving rise to electromagnetic interactions, these are attractive when the vibrational frequencies and absorptions are identical or similar, repulsive when non-identical. Other interactions originally proposed by van der Waals were included in this name, but these are usually separated into the Coulomb's force, the Keesom force and the London force. Only the last is of electrodynamic nature. Probably it is important for holding lipid membranes into that structure and possibly in other interactions, for example cell adhesion. Electrodynamic forces between large-scale assemblies can be of relatively long range nature. (Source: *Dictionary of Cell and Molecular Biology*)

    van der Waals forces include momentary attractions between molecules, diatomic free elements, and individual atoms. They differ from covalent and ionic bonding in that they are not stable, but are caused by momentary polarization of particles.

3. Kim Eric Drexler   An American Engineer best known for popularizing the potential of molecular nanotechnology (MNT), in the 1970s and 1980s. His doctoral thesis at MIT was revised and published as the book *Nanosystems: Molecular Machinery, Manufacturing and Computation* (1992) which recieved the Association of American Publishers award for best computer science book of 1992. He also coined the term Grey Goo.

4. N. Taniguchi, "On the Basic Concept of Nano-Technology," Proc. Intl. Conf. Prod. Eng. Tokyo, Part II, Japan Society of Precision Engineering, 1974.

5. The First Discovery of Fullerenes C_{60} Buckminsterfullerene   This was the experiment performed by three Nobel Prize Winners in 1985, Robert. F. Curl, Sir Harold W. Kroto and Richard E. Smalley. Their experiments aimed at understanding the mechanisms by which long chained carbon molecules are formed in interstellar space and circumstellar shells, graphite was vaporized by laser irradiation, producing a remarkably stable cluster consisting of 60 carbon atoms. Concerning the question of what kind of 60-carbon atom structure might give rise to a superstable species, they suggested a truncated icosahedron, a polygon with 60 vertices and 32 faces, 12 of which are pentagonal and 20 hexagonal. This object was commonly encountered as the football. The C_{60} molecule which resulted when a carbon atom was placed at each vertex of this structure had all valences satisfied by two single bonds and one double bond, had many resonance structures, and appeared to be aromatic.

6. Deterministic approaches in which single molecules (created by stochastic chemistry) are manipulated on substrate surfaces (created by stochastic deposition methods) by deterministic methods comprising nudging them with STM or AFM probes and causing simple binding or cleavage reactions to occur.

7. Microelectromechanical systems (MEMS)   It is the technology of the very small and merges at the nanoscale into nano-electromechanical systems (NEMS) and nanotechnology. MEMS are also called as micromachines (in Japan), or *Micro Systems Technology—MST* (in Europe). MEMS are separate and distinct from molecular electronics, the hypothetical vision of nanotechnology. MEMS are made up of components between 1 to 100 nm in size (i.e., 0.001 to 0.1 mm). MEMS devices generally range in size from 20 μm (20 millionth of a metre) to a millimetre (thousandth of a metre). They usually consist of a central unit that processes the data, the microprocessor and several components that interact with the outside such as microsensors. At these size scales, the standard principles of Physics are not applicable. Due to MEMS' large surfac-area-to-volume ratio, surface effects such as electrostatics and wetting dominate volume effects such as inertia or thermal mass.

REVIEW QUESTIONS

1. Write short notes on pre-nanotechnology.

2. Mention the role of important scientists in the development of the "Nano" concept.

3. Explain Moore's law.

4. Define nanometre.

5. Define nanotechnology.

6. State the stages of development of nanotechnology.

7. What is general-purpose technology?

8. Explain multi-purpose technology.

9. Define exponential technology.

10. Brief the emerging fields of nanotechnology.

11. Discuss in detail the applications of nanotechnology.

12. Write short notes on applications of nanotechnology in

    i.   nanomedicine

    ii.   chemistry and environment

    iii. energy

iv.   information and communication

v.   heavy industry

vi.   consumer goods

REFERENCES

1. Gall, John, (1986). *Systemantics: How Systems Really Work and How They Fail*, 2nd edn. The General Systemantics Press, Ann Arbor, MI.

2. Hillie Thembela and Hlophe Mbhuti. (2007). "Nanotechnology and the challenge of clean water." *Nature Nanotechnology.* 2: 663–664.

3. Kahn, Jennifer (2006). "Nanotechnology." *National Geographic.* June Issue. 98–119.

4. Waldner, Jean-Baptiste (2007). *Nanocomputers and Swarm Intelligence.* ISTE, London p.26.

5. Zsigmondy, R. (1914). *Colloids and the Ultramicroscope.* J.Wiley and Sons, NY.

Chapter 2

Nanochemistry

INTRODUCTION

Nanochemistry is a scientific branch of nanotechnology or nanoscience that examines the synthesis, analysis and structure of chemical compounds at the nanoscale. Nanochemistry is a new discipline concerned with the unique properties associated with assemblies of atoms or molecules on a scale between that of the individual building blocks and the bulk material. At this level, quantum effects can be significant, and also innovative ways of carrying out chemical reactions become possible.

Nanochemistry is the use of synthetic chemistry to make nanoscale building blocks of desired shape, size, composition and surface structure, charge and functionality with an optional target to control self-assembly of these building blocks at various scale-lengths.

Nanochemistry, in particular, presents a unique approach to building devices with a molecular-scale precision. The main challenges to full utilization of nanochemistry centre on understanding new rules of behaviour, because nanoscale systems lie at the threshold between classical and quantum behaviour and exhibit behaviours that do not exist in larger devices.

Although nanochemical control was proposed decades ago, it was only recently that many of the tools necessary for studying the nanoworld were developed. These include the scanning tunnelling microscope (STM), atomic force microscope (AFM), high resolution scanning and transmission electron microscopes, X-rays, ion and electron beam probes, and new methods for nanofabrication and lithography.

Studies of nanochemical systems span many areas, from the study of the interactions of individual atoms and how to manipulate them, how to control chemical reactions at an atomic level, to the study of larger molecular assemblies, such as dendrimers, clusters, and polymers. From studies of assemblies, significant new structures—such as nanotubes, nanowires, three-dimensional molecular assemblies, and lab-on-a-chip devices for separations and biological research—have been developed.

This chapter deals with nanomaterials, fullerenes, nanoparticles, carbon nanotube and its related structures, colloidal gold, nanostructures, nanoscale iron particles and self-assembled monolayers which are very much important areas of study, research and development in the nanochemistry field. The area of nanochemistry is getting wider day by day. Important areas are dealt with in this chapter.

BASIC CONCEPTS

Nanomaterials are materials which are derived from the processes of nanotechnology and used in nanotechnological applications. They have got the important structural criteria, namely, the size. It is smaller than one-tenth of a micrometre at least in one dimension. No consensus upon the minimum or maximum size of nanomaterials is arrived at, at the moment. Their size ranges from 1 to 30 nm, i.e., the nanoscale of nanomaterial is defined as between microscale (0.1 mm) and atomic/molecular scale (about 0.2 nm).

When the nanoscale materials are studied, their unique property is the vastly increased ratio of surface-area-to-volume present in many nanoscale materials. This creates new quantum mechanical effects like that of the "quantum size effect" where the electronic properties of solids are altered with great reductions in particle size. This effect is not due to the size reduction from macro- to microdimensions. These properties are well brought out when the nanometre size range is reached for any particular material. Various physical properties are changing compared to macroscopic systems, when the materials are in nanoscale. Nanomechanics research is concentrating on the novel mechanical properties of nanomaterials. Catalytic activity of nanomaterials reveals peculiar properties in the interaction with biomaterials.

Nanotechnology can be considered as extensions of traditional disciplines from physics to biology for utilizing the properties derived from the nanosize of the materials. Traditional disciplines can be reinvented in terms of specific applications of nanotechnology. This sort of interchanging concepts and ideas among macro- and nano- contributes to the modern understanding of the new and emerging field of science of nano. In a broader outlook, nanotechnology is the synthesis and application of ideas from science and engineering towards the understanding and production of novel materials and peculiar devices. These products generally make use of physical properties associated with small scales, extensively.

Materials produced at nanoscale have got very different properties compared to what materials on a macroscale are having and so they have unique applications than macromaterials. Some examples are: opaque substances become transparent (copper); inert materials become catalysts (platinum); stable materials turn combustible (aluminium); solids turn into liquids at room temperature (gold); insulators become conductors (silicon). Gold, as a material is chemically inert at normal scales but develops the properties which can be used as a potent chemical catalyst at nanoscales. These unique quantum and surface phenomena that matter exhibits at the nanoscale triggers the interest and curiosity and make more researchers to be attracted towards nanotechnology.

Nanosize powder particles (also called as nanoparticles which are few nanometres in diameter) are potentially important in ceramics, powder metallurgy, areas which require the attainment of uniform nanoporosity and similar applications. The strong tendency for the formation of clumps (agglomerates) while using such small particles is a difficult-to-solve technological problem that prevents utilizing them in such applications. A few dispersants such as ammonium citrate (aqueous) and imidazoline or oleyl alcohol (non-aqueous) are promising additives for deagglomeration in such situations to avoid agglomerates in processes requiring nanoparticles.

Nanomaterials have got different behaviours than other similar-sized particles. Specialized approaches are needed for testing and monitoring their effects on human health and on the environment and are to be developed necessarily.

Nanomaterials and nanotechnologies are projected to yield numerous health and health care advances, such as more targeted methods of delivering drugs, new cancer therapies and methods of early detection of diseases. They have possible unwanted effects. Increased rate of absorption is the main concern associated with manufactured nanoparticles related to unwanted effects on health.

Very high surface-area-to-volume ratio, the important property of nanoparticles, otherwise called as the greater specific surface area (surface area per unit weight) may lead to increased rate of absorption through the skin, lungs or digestive tract and may cause unwanted effects on the lungs as well as other organs. The particles must be absorbed in sufficient quantities in order to pose health risks is an important point to be noted.

CLASSIFICATION OF NANOMATERIALS

The study of the nanomaterials are divided into following subfields.

1. Carbon nanotubes and other fullerenes and various nanoparticles and nanorods are materials derived from the interface and colloid science which have potential utility in nanotechnology.

2. Nanoscale materials can be used for bulk applications and most of the present applications of nanotechnology in commercial way are of this nature.

3. Progress has been made in using these materials for medical applications—nanomedicine.

Techniques of Production or Methodology

Bottom-up methodology   Bottom-up approach is one in which the required nanomaterials are arranged from smaller components, i.e., smaller components into more complex assemblies. It is like building a wall by using smaller bricks and laying a road by using small stones.

1.   DNA nanotechnology utilizes the specificity of Watson–Crick base-pairing to construct well-defined structures out of DNA and other nucleic acids.

2.   More generally, molecular self-assembly seeks to use concepts of supramolecular chemistry and molecular recognition in particular, to cause single-molecule components to automatically arrange themselves into some useful conformation.

Top-down methodology   Top-down approach seeks to create smaller devices by using larger ones to direct their assembly. The big stone being made into pieces can be compared to this method. It is like producing stones from mountains.

Many technologies that have descended from conventional solid-state silicon methods for fabricating microprocessors which are now capable of creating features smaller than 100 nm, fall under the definition of nanotechnology. Examples are giant magnetoresistance-based hard drives and atomic layer deposition (ALD) techniques. Solid-state techniques can also be used to create devices known as nanoelectromechanical systems (NEMS), which are related to microelectromechanical systems (MEMS).

Atomic force microscope tips can be used as a nanoscale "write head" to deposit a chemical upon a surface in a desired pattern in a process called dip pen nanolithography. This fits into the larger subfield of nanolithography.

Functional methodology   In this functional approach, the developed materials will have desired functionality without regard to how they are assembled like bottom-up or top-down approaches.

1.   Molecular electronics is a field in which the development of molecules with useful electronic properties are envisaged. These molecules could then be used as single-molecule components in a nanoelectronic device like rotaxane. In this methodology, the developed materials have got specific functions to perform.

2.   Synthetic chemical methods can be used to create synthetic molecular motors which are useful in nanocar-like devices.

Speculative methodology   The speculative or dreaming or imaginary methodologies are the ones in which futuristic way of expecting inventions in nanotechnology and the way in which the field of nanotechnology will progress in future and the proposals for the expansion of nanotechnology are drawn theoretically. These fields create a broader view of nanotechnology with more emphasis on its societal implications than the details of creations of such inventions actually in practice.

SIZE CONCERNS

An important point related to matter or materials science is that the volume of an object decreases as the third power of its linear dimensions, but the surface area only decreases as its second power. This subtle and unavoidable principle due to miniaturization of materials into nanoscale has huge ramifications. If a functionality of a drill is taken into consideration using the above principle, the power of a drill (or any other machine) is proportional to the volume, while the friction of the drill's bearings and gears is proportional to their surface area. For a normal-sized drill, the power of the device is enough to overcome any friction but smaller drills will have 10 times as much friction as power. In that case, the nanodrill is useless.

Super-miniature electronic integrated circuits are possible to function but in reality the same technology cannot be used to make functional mechanical devices in miniature; the friction developing out of nanoscale level overtakes the available power at such small scales. Microphotographs of delicately etched silicon gears created out of speculative methodology which are seen in articles are curiosities with limited real-world applications. In moving mirrors and shutters, surface tension increases in the same way, causing very small objects having the tendency to stick together. This sort of same surface tension increase could possibly make any kind of "microfactory" impractical: robotic arms and hands of nanoscale even if created, whatever they pick up will tend to be impossible to be separated.

In spite of the above difficulties related to mechanical factors, molecular evolution has produced the working cilia, flagella, muscle fibres and rotary motors in aqueous environments, all on the nanoscale in the biological system. These machines exploit the increase of the frictional forces found at the micro- or nanoscale. Unlike the machine parts like an oar, paddle or propeller, the mechanics of which are dominated by normal frictional forces (the frictional forces perpendicular to the surface) for propulsion, cilia and other molecular evolution products develop motion resulting from the exaggerated drag or laminar forces (frictional forces parallel to the surface) present at micro- and nanodimensions.

For creating meaningful "machines" at the nanoscale, the relevant forces which act in favour and against need to be considered. The simple reproductions of macroscopic ones are not the expected thing in nanotechnology but the development and design of relevant machines. It is really challenging at the moment to create such machines.

Materials that come under the purview of nanomaterials are divided into two categories: 1) Fullerenes 2) Inorganic nanoparticles.

FULLERENES

INTRODUCTION

Fullerenes are a form of nanomaterial, belonging to a family of carbon allotropes. They are molecules that are composed entirely of carbon. They may take the shape of a hollow sphere, ellipsoid, tube, or plane. Spherical fullerenes are named as buckyballs and cylindrical ones are called as carbon nanotubes or buckytubes. Graphene is an example of a planar fullerene sheet. Fullerenes are similar in structure to graphite. Usually, graphite is composed of stacked sheets of linked hexagonal rings, but may also contain pentagonal or heptagonal rings that would prevent a sheet from being planar. They are of interest in nanotechnology due to both their tensile strength and their electrical properties.

HISTORY OF DISCOVERY

In 1985, Robert Curl, Harold Kroto and Richard Smalley at the University of Sussex and Rice University discovered fullerenes. They are named after Richard Buckminster Fuller.

Kroto, Curl, and Smalley were awarded Nobel Prize in chemistry for their roles in the discovery of fullerenes in 1996. C_{60} and other fullerenes occurring in other form of materials as in normal candle soot were noticed later. By 1991, gram-sized samples of fullerene powder were easily produced but the purification remains a challenge to chemists. Fullerene is an unusual reactant in many organic reactions. The buckminsterfullerenes, in the form of C_{60}, C_{70}, C_{76}, and C_{84} molecules, are produced in nature in minute quantities. They are hidden in soot and also formed by lightning discharges in the atmosphere. Recently, a family of minerals known as Shungites in Karelia, Russia was found to contain buckminsterfullerenes.

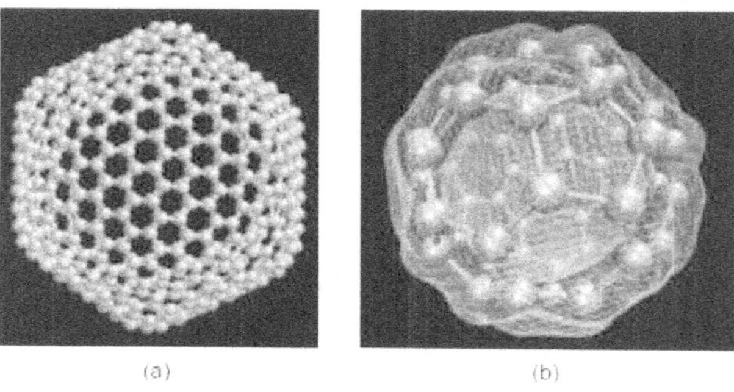

Figure 2.1 The isosahedral fullerenes (a) C_{540} (b) C_{60}

In 1970, Eiji Osawa of Toyohashi University of Technology predicted the existence of C_{60} and this was reported in Japanese magazines, but not in Europe or America. The structure of a corannulene[1] molecule that was a subset of a soccer-ball shape was noticed by him and a hypothesis that a full ball shape could also exist was made by him. Figure 2.1 shos the structure of fullerenes C_{540} and C_{60}.

VARIATIONS

From 1985, after the discovery of fullerenes, structural variations have evolved well beyond the individual clusters themselves. Examples of fullerenes are as follows:

Buckyball clusters   Smallest member is C_{20} (unsaturated version of dodecahedrane) and the most common is C_{60}. Buckminsterfullerene, the smallest fullerene molecule of the fullerene family, has the shape of a pentagon. It can often be found in soot and is also the most common in terms of natural occurrenc. It is also known as buckyball (Figure 2.2).

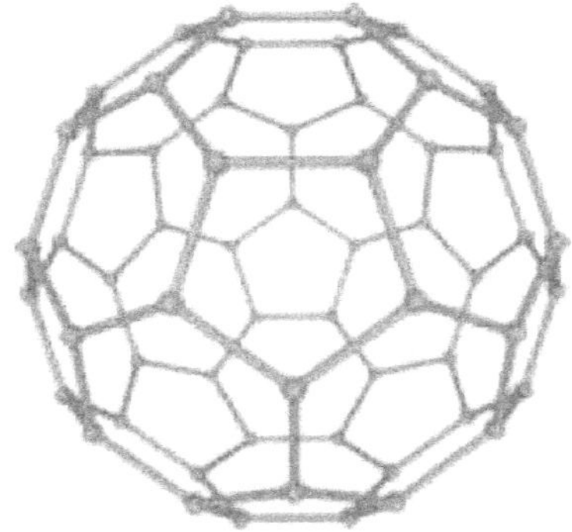

Figure 2.2   Buckminsterfullerene C_{60}

Buckyball utilizing boron atoms instead of the usual carbon as a new type has been predicted and described by researchers at Rice University. These boron bucky ball (B-80 structure) is predicted to be more stable than the C_{60} buckyball. The reason is that the B-80 is actually more like the original geodesic dome structure popularized by Buckminster Fuller which utilizes triangles rather than hexagons.

Nanotubes   Hollow tubes of very small dimensions, having single or multiple walls; They have potential applications in electronics industry. Nanotubes are a form of nanomaterial and are nothing but cylindrical fullerenes. These tubes of carbon are usually a few nanometres wide, but they can range from less than a micrometre to several millimetres in length. They often have closed ends, but can be open-ended as well. There are also cases in which the tube reduces in diameter before closing off.

Nanobuds   Nanobuds are structures derived by adding buckminsterfullerenes to carbon nanotubes.

Megatubes   Larger in diameter than nanotubes and prepared with walls of different thickness; potentially used for the transport of a variety of molecules of different sizes.

Polymers   In the form of chain, two-dimensional and three-dimensional polymers are formed under high pressure and high temperature conditions.

Nano "onions"   Spherical particles based on multiple carbon layers surrounding a buckyball core; proposed for lubricants.

Linked "ball-and-chain" dimers   Two buckyballs linked by a carbon chain.

Fullerene ring assemblies   Two identical fullerenes or modified fullerenes joined by a single or double bond can be named by ring assembly nomenclature. Two different fullerenes are named by substituting one fullerene structure by the other.

PROPERTIES OF FULLERENES

The chemical and physical properties of fullerenes have been a hot topic in the field of research and development during the past decade. The same interest is likely to continue in future years. The possible uses of fullerenes in armour have been published in *Popular Science*. Fullerenes were researched for potential medicinal use like binding specific antibiotics to the structure to target resistant bacteria and even target certain cancer cells such as melanomas. An article describing the use of fullerenes as light-activated antimicrobial agents is found in *Chemistry and Biology Journal*.

In the field of nanotechnology, heat resistance and superconductivity of these materials are some of the more heavily studied properties.

A common method of producing fullerenes is to supply more quantity of current between two nearby graphite electrodes in an inert atmosphere. The resulting carbon plasma arc between the electrodes is the result. It cools into sooty residue from which many fullerenes can be isolated.

Aromaticity

Increasing the reactivity of fullerenes is possible by attaching active groups to their surfaces. Buckminsterfullerene does not exhibit "superaromaticity" (i.e., the electrons in the hexagonal rings do not delocalize over the whole molecule).

Chemistry

Fullerenes are stable and also reactive at times. Endohedral fullerenes are inclusion compounds which are produced by trapping the other atoms inside the fullerenes.

An unusual example is the egg-shaped fullerene $Tb_3N@C_{84}$. This is the one that does not conform to the isolated pentagon rule. Recent evidence for a meteor impact at the end of the Permian period was found. It was concluded by analysing noble gases so preserved. Metallofullerene-based inoculates using the rhonditic steel processes are the production methodology put into practical use and one of the first commercially viable uses of buckyballs.

Solubility

In many solvents, fullerenes are sparingly soluble. Common solvents such as aromatics (toluene) and others like carbon disulphide are used. Fullerenes have a unique property in that they are the only known allotropes of carbon that can be dissolved in common solvents at room temperature.

Some fullerene structures are insoluble due to a small band gap between the ground and excited states. The small fullerenes C_{28}, C_{36} and C_{50} are the examples. Small band gap fullerenes are highly reactive in nature and bind to other fullerenes or to soot particles easily. Solubility of C_{60} in some solvents shows unusual behaviour. It is due to the existence of solvate phases or analogues of crystallohydrates.

Quantum Mechanics

Molecules such as fullerenes have got the wave–particle duality and it was demonstrated in 1999 by researchers from the University of Vienna. Julian Voss-Andreae[2] one of the co-authors of this research, became an artist. He has since created several sculptures symbolizing wave-particle duality in buckminsterfullerenes.

Safety and Toxicity

Fullerenes (C_{60}, C_{70},...) and fullerene derivatives differ in their toxicological data. C_{60} or other fullerenes are coming under covalently bonded chemical groups, whereas fullerene complexes (e.g. C_{60}-PVP, host–guest complexes) are the ones in which the fullerene is physically bound to another molecule.

The above different compounds are in the range of insoluble materials in either hydrophilic or lipophilic media or even amphiphilic compounds. Also they have got other varying physical and chemical properties. Therefore, any broad generalization extrapolating or grouping them under single heading is not possible.

There is no definite evidence to show that C_{60} is toxic while reviewing literature on fullerene toxicity starting from the early 1990s to present day. Research shows that results for C_{60} and nanotubes are having no discrepancy from a toxicological standpoint.

NANOPARTICLES

INTRODUCTION

Nanoparticles or nanocrystals derived from metals, semiconductors or oxides are of interest for their mechanical, electrical, magnetic, optical, chemical and other properties in various applications. Nanoparticles have been used routinely as quantum dots and as chemical catalysts in various commercial applications.

Nanoparticle is the particle described and used in nanotechnology. A particle is defined as a small object behaving as a whole unit in terms of its transport and properties. Particle is classified according to their size as fine and ultrafine particles. Fine particles have got a diameter range of 100–2500 nm. Ultrafine particles have got a diameter range of 1 and 100 nm. Nanoparticles are like ultrafine particles sized between 1 and 100 nm, though the size limitation can be restricted to two dimensions. Nanoparticles may or may not have size-related intensive properties. The properties of nanoparticles are different from that of fine particles or bulk materials in a significant way.

Nanoparticles have greater scientific applications as they are effectively a bridge between bulk materials and atomic or molecular structures. A bulk material is the one which has constant physical properties regardless of its size. The same is not applicable at nanoscale. Some of the size-dependent properties are observed, e.g. quantum confinement in semiconductor particles, surface plasmon resonance in some metal particles and supermagnetism in magnetic materials.

Nanoparticles have a very high surface-area-to-volume ratio. This ratio is responsible for tremendous driving force for diffusion, especially at elevated temperatures. Sintering[3] can take place at lower temperatures, over shorter time scales than for larger particles. These changing temperatures have no effect on the density of the final products theoretically. But it can produce flow difficulties, and change the tendency of nanoparticles to agglomerate complicated matters and ultimately the final product.

Nanoclusters are the form of nanoparticles and they have at least one dimension between 1 and 10 nm and a narrow size distribution. Nanopowders (Figure 2.3) are agglomerates of ultrafine particles, nanoparticles, or nanoclusters. Nanocrystals are nanometre-sized single crystals or single-domain ultrafine particles. Nanoparticle research is an area of intense scientific research currently due to a wide variety of potential applications in biomedical, optical, nd electronic fields. Table 2.1 lists the various nanoparticles.

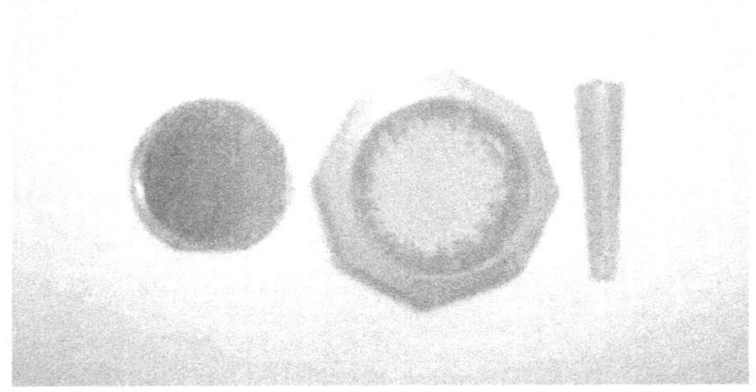

Figure 2.3 Silicon nanopowder

Table 2.1 Classification of nanoparticles

Particle	Size
Coarse particle	Less than 10 microns
Fine particle	100–2500 nm (less than 2.5 microns)
Ultrafine particle	1–100 nm (less than 0.1 microns = 100 nm)
Nanoparticle	1–100 nm
Nanocluster	1–10 nm
Nanopowder	Agglomerates of ultrafine particles, nanoparticles or nanoclusters

HISTORY

Nanoparticles were known to the human civilization long ago. Generally they are considered as an invention of modern science because of its recent description in terms of material science. In Mesopotamia, nanoparticles were specifically used by artisans for generating a glittering effect on the surface of pottery even during the 9th century AD.

Lustre Technique

Even now, analysis of pottery from the middle ages and renaissance show a distinct gold or copper-coloured metallic glitter. This is retained for so many centuries or years. This lustre is due to a metallic film that was applied to the transparent surface of a glazing. The lustre is visible even today, provided the film has resisted atmospheric oxidation and other weathering. The origin of lustre is due to silver and copper nanoparticles, dispersed homogeneously in the glassy matrix of the ceramic glaze contained within the film. The lustre-creating nanoparticles were produced by the artisans by adding copper and silver salts and oxides together with vinegar, ochre, and clay, on the surface of previously glazed pottery. The object was heated to about 600°C in a reducing atmosphere after placing in a kiln. Due to the heat, the glaze would soften. The softening was responsible for the migration of copper and silver ions into the outer layers of the glaze. Reduction of the ions back to metals occurs on the outer layer because of the reducing atmosphere. The metals thus reduced come together forming the nanoparticles that give the colour and optical effects.

Lustre technique gives us an idea that craftsmen of yesteryears had a rather sophisticated empirical knowledge of materials. The technique was originated in the Islamic world because of the use of gold in artistic representations. So, this way was adopted to create a similar effect without using real gold. The solution was found in lustre.

The first description, in scientific terms, of the optical properties of nanometre-scale metals was provided in "Experimental relations of gold (and other metals) to light," by Michael Faraday in the classic paper of 1857.

PROPERTIES OF NANOPARTICLES

Nanoparticles effectively form a bridge between bulk materials and atomic or molecular structures. So, they are of great scientific interest. A bulk material should have constant physical properties regardless of its size. At the nanoscale this is not applicable. When the bulk material is made into nanomaterial, size-dependent properties are observed. They are quantum confinement in semiconductor particles, surface plasmon resonance in some metal particles and superparamagnetism in magnetic materials.

The properties of materials change at the nanoscale and the percentage of atoms at the surface of a material also becomes significant. The percentage of atoms at the surface is minuscule relative to the total number of atoms of the material related to bulk materials larger than one micrometre. The interesting and unexpected properties of nanoparticles are partly due to the uniqueness of the surface of the material compared to that of the bulk materials.

Nanoparticles show a number of special properties relative to bulk material. Taking copper as an example, this can be explained. The bending of bulk copper (wire, ribbon, etc.) occurs with movement of copper atoms/clusters at about the 50 nm scale. Copper nanoparticles smaller than 50 nm reach the level of superhard materials. They do not exhibit the same malleability and ductility as bulk copper. This sort of change in properties is not desirable always. Ferroelectric materials smaller than 10 nm have got the property of switching their magnetization direction using room-temperature thermal energy. This change makes them useless for memory storage. Suspensions of nanoparticles are possible as the interaction of the particle surface with the solvent is strong enough to overcome differences in density. This difference in density usually results in a material either sinking or floating in a liquid. Nanoparticles often have unexpected visible properties due to nanosize which confine their electrons and produce quantum effects. For example, gold nanoparticles appear deep red to black in solution compared to the yellow colour with which we ssociate gold always. Figure 2.4 shows the TEM image of nanodiamonds.

Figure 2.4 Nanodiamonds (TEM image)

Nanoparticles are found to impart different properties to various day-to-day products. Titanium dioxide nanoparticles impart the self-cleaning effect and the size being nanorange, the particles cannot be seen. Nano zinc oxide particles are found to have superior UV blocking properties compared to its bulk substitute. So, it is used often in the sunscreen lotions. Clay nanoparticles increase reinforcement, leading to stronger plastics when incorporated into polymer matrices. This is verified by a higher glass transition temperature and other mechanical property tests. These nanoparticles are hard, and impart their properties to the polymer (plastic). In order to create smart and functional clothing, nanoparticles are attached to textile fibres.

CLASSIFICATION

Nanoparticles are classified based on their size, shape and utility.

- ❖ By size   Nanoclusters at the small end of size range
- ❖ By shape   Nanospheres, nanorods and nanocups
- ❖ By utility   This may be discussed as under:

     1.   Metal, dielectric and semiconductor nanoparticles and hybrid structures (e.g. core-shell nanoparticles).

2. *Quantum dots*   Nanoparticles are made of semiconducting material that is labelled as quantum dots if sub 10 nm, and quantization of electronic energy levels occurs. They are used in biomedical applications as drug carriers or imaging agents.

3. *Semi-solid and soft nanoparticles*   A prototype nanoparticle of semi-solid nature is the liposome. Various types of liposome nanoparticles are currently used clinically as delivery systems for anti-cancer drugs and vaccines.

CHARACTERIZATION

Nanoparticle characterization is essential to establish the understanding and control of nanoparticle synthesis and applications. Characterization is done by a variety of different techniques, which are from materials science. Common techniques are as follows:

1. Electron microscopy (TEM, SEM)

2. Atomic force microscopy (AFM)

3. Dynamic light scattering (DLS)

4. X-ray photoelectron spectroscopy (XPS)

5. Powder X-ray diffractometry (XRD)

6. Fourier transform infrared spectroscopy (FTIR)

7. Matrix-assisted laser-desorption time-of-flight mass spectrometry (MALDI-TOF)

8. Ultraviolet-visible spectroscopy.

Brownian movement (after Robert Brown) has been known for over a century. The technology for nanoparticle tracking analysis (NTA) allows direct tracking of the motion. This method also allows the sizing of individual nanoparticles in solution.

PRODUCTION OF NANOPARTICLES

There are several methods for creating nanoparticles. Attrition and pyrolysis are common methods.

1. In attrition, macro- or microscale particles are ground in a ball mill, a planetary ball mill or other size-reducing mechanism. The resulting particles are air-classified to recover nanoparticles.

2. In pyrolysis, an organic precursor (liquid or gas) is forced through an orifice at high pressure and burned. The resulting ash is air-classified to recover oxide nanoparticle.

3. Thermal plasma can also be used as an energy source necessary to cause evaporation of small micrometre size particles. The thermal plasma temperatures are in the order of 10000 K, so that the solid powder easily evaporates. Nanoparticles are formed upon cooling while exiting the plasma region. The radio frequency (RF) plasma method has been used to synthesize different nanoparticle materials. Synthesis of various ceramic nanoparticles such as oxides, carbides and nitrides of Ti and Si are done using this method.

4. Inert-gas aggregation is frequently used to make nanoparticles from metals with low melting points. The metal is vaporized in a vacuum chamber and then supercooled with an inert-gas stream. The super-cooled metal vapour condenses into nanometre-sized particles. That can be entrained in the inert-gas stream and deposited on a substrate or studied *in situ.*

MORPHOLOGY OF NANOPARTICLES

Naming these particles after the real shapes like nanospheres, nanoreefs, nanoboxes, etc. is the rule followed (Figure 2.5). These morphologies are of spontaneous origin due to the effect of templating or directing agent present in the synthesis such as micellar emulsions or anodized alumina pores. It may be due to the innate crystallographic growth patterns of the materials themselves. These morphologies may have a purpose, like long carbon nanotubes being used to bridge anelectrical junction, or just a scientific curiosity like the stars shown in Figure 2.5.

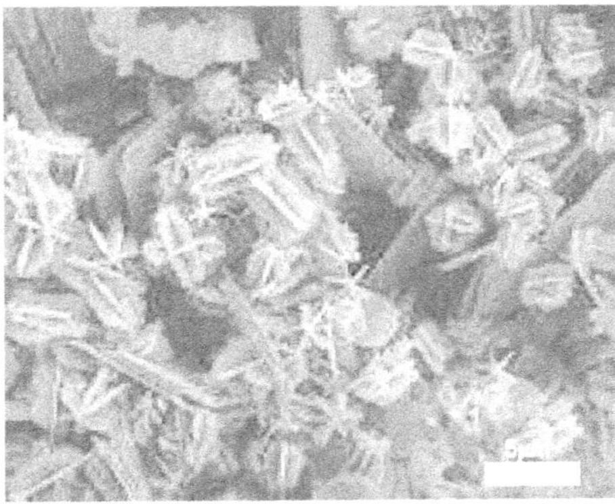

Figure 2.5 Nanostars of vanadium (IV) oxide

SAFETY ISSUES

Nanoparticles present possible dangers, both medically and environmentally. Most of the dangers are due to the high surface-area-to-volume ratio, which can make the particles very reactive or catalytic. They are also able to pass through cell membranes in organisms, and their interactions with biological systems are relatively unknown. Free nanoparticles in the environment quickly tend to agglomerate and leave the nano-regime. Nature itself would have evolved immunity for organisms on earth to many nanoparticles such as salt particulates from ocean aerosols, terpenes from plants or dust from volcanic eruptions.

Animal studies have shown that some nanoparticles can penetrate cells and tissues, move through the body and brain and cause biochemical damage. Cosmetics and sunscreens containing nanomaterials are posing health risks and its direct impact remains largely unknown, pending completion of long-range studies. Diesel nanoparticles have been found to damage the cardiovascular system in a mouse model.

CARBON NANOTUBE

INTRODUCTION

Carbon nanotubes (CNTs) belong to the family of nanomaterials. They are allotropes of carbon with a nanostructure that can have a length-to-diameter ratio greater than 1,000,000 and as high as 40,000,000. They are cylindrical carbon molecules. Their novel properties make them potentially useful in many applications like in nanotechnology, electronics, optics and other fields of materials science and also have got extensive use in **arcology**[4] and architectural fields. They have got extraordinary strength, unique electrical properties and are efficient conductors of heat. Apart from organic nanotubes, inorganic nanotubes have also been ynthesized. The three-dimensional structure of a carbon nanotube is shown in Figure 2.6.

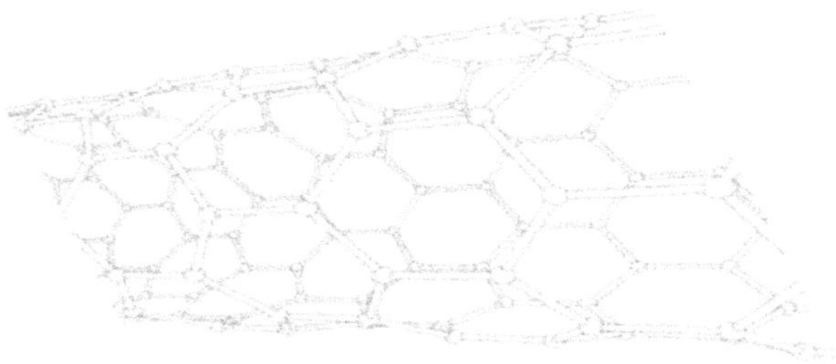

Figure 2.6   3D structure of a carbon nanotube

DISCOVERY

In Soviet *Journal of Physical Chemistry*, L. V. Radushkevich and V. M. Lukyanovich published clear images of 50-nm-diameter tubes made of carbon in 1952. In 1991,

Sumio Iijima of NEC has been attributed with the discovery of hollow, nanometre-sized tubes composed of graphitic carbon. The invention of the transmission electron microscope (TEM) allowed the direct visualization of nanotube structures. A U.S. patent for the production of "cylindrical discrete carbon fibrils" was issued to Howard G. Tennent of Hyperion Catalysis in 1987. But, it was Iijima's discovery of carbon nanotubes in the insoluble material of arc-burned graphite rods that attracted the attention of the world towards the carbon nanotubes. Professor Tang Zikang and Wang Ning were responsible for creating the smallest stable carbon nanotubes in the world, measuring just 0.4 nm in diameter, in 2000. Nanotube transistor can be grown in bulk, like that of silicon transistors and this was demonstrated in 2001 by IBM researchers. The process utilized by them is called "constructive destruction". This includes the automatic destruction of defective nanotubes on the wafer.

Single-chip wafers with over ten billion correctly aligned nanotube junctions have been created following this development. The removal of incorrectly aligned nanotubes automatically using standard photolithography equipment has also been demonstrated.

In 2004, the first nanotube integrated memory circuit was created. The conductivity of nanotubes regulation is one of the main challenges. A nanotube may act as a plain conductor or as a semiconductor which depends on subtle surface features. Removal of non-semiconductor tubes using a fully automated method has been developed.

Random networks of carbon nanotubes have been used to make transistors alternatively. This makes their electrical differences being averaged and that allows production of devices on a large scale at the wafer level.

DESCRIPTION

Nanotubes are part of the fullerene structural family. Spherical buckyballs also belong to the same family. The cylindrical nanotube usually has at least one end capped with a hemisphere of the buckyball structure, the other end being open. The name nanotube is because of their size. The diameter of a nanotube is in the order of a few nanometres (approximately 1/50,000th of the width of a human hair), but their length can be up to several millimetres.

In applied quantum chemistry, specifically, orbital hybridization describes about the nature of the bonding of a nanotube. van der Waals forces allow nanotubes to align themselves into bundles of ropes naturally. Under high pressure, nanotubes can merge together, by exchange of bonds. It gives the possibility of producing strong, unlimited-length wires through high-pressure nanotube linking.

Their unique molecular structure results in extraordinary macroscopic properties, including high tensile strength, high electrical conductivity, high ductility, high resistance to heat and relative chemical inactivity (as it is cylindrical and "planar"—that is, it has no "exposed" atoms that can be easily displaced). Carbon nanotubes are used in paper batteries, developed by researchers at Rensselaer Polytechnic Institute and another one is in the field of space technologies and is used to produce high-tensile carbon cables required by a space elevator.

TYPES OF CARBON NANOTUBES AND RELATED STRUCTURES

Carbon nanotubes come under the following categories.

1. Single-walled (SWNT) nanotubes
2. Multi-walled (MWNT) nanotubes
3. Fullerite
4. Torus
5. Nanobud
6. Nanoflower

Single-walled Nanotubes (SWNT)

Mostly single-walled nanotubes have a diameter of close to 1 nm. The tube length can be many thousands of times longer than the diameter. The structure of a SWNT can be visualized by wrapping a one-atom-thick layer of graphite called graphene into a seamless cylinder. SWNT exhibits important electrical properties which are not shared by the multi-walled carbon nanotube (MWNT) variants. Single-walled nanotubes are the pathway for miniaturizing electronics beyond the current basis of modern electronics, namely, the microelectromechanical scale. The most basic building block of these systems is the electric wire and SWNTs can be excellent conductors. Useful applications of SWNTs are in the development and production of the first intramolecular field effect transistors (FETs) and in turn the production of the first intramolecular logic gate using SWNT FETs. But, this process is very expensive. The development of more affordable synthesis techniques is vital to the future of carbon nanotechnology.

Multi-walled Nanotubes (MWNT)

Multi-walled nanotubes are the ones in which multiple layers of graphite are rolled in on themselves to form a tube shape. The structures of multi-walled nanotubes can be described using two models. 1) Russian doll model and 2) Parchment model. The special property of double-walled carbon nanotubes (DWNT) is that, they combine very similar morphology and properties as compared to SWNT, while their resistance to chemicals improved significantly.

Fullerite

Fullerites (Figure 2.7) are the solid-state form of fullerenes and related compounds and materials. Polymerized single-walled nanotubes (P-SWNT), being highly incompressible nanotube forms, are a class of fullerites and can be compared to diamond in terms of hardness. Because of the way the nanotubes intertwine, P-SWNTs structurally have no corresponding crystal lattice as that of diamonds. That makes it impossible to cut fullerite like the diamonds neatly. This same structure resuls in a less brittle material. So, any impact that the structure sustains will spread throughout the material.

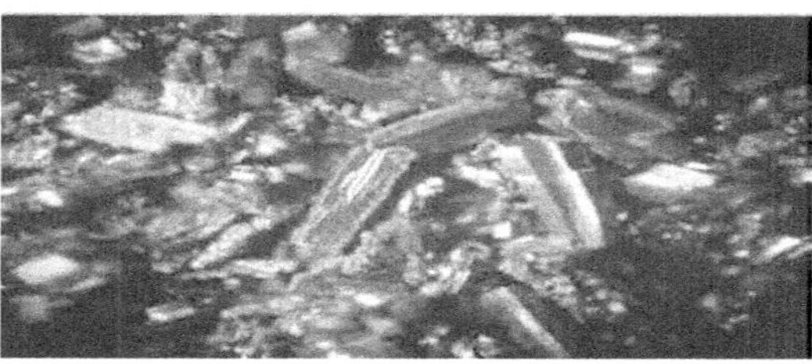

Figure 2.7   Fullerites

Ultrahard fullerite   Ultrahard fullerite (C_{60}) is a form of carbon, synthesized under high pressure and high temperature conditions. It is believed that the fullerene molecules are three-dimensionally polymerized in this material.

Torus

A nanotorus is a theoretically described carbon nanotube which is bent into a torus (doughnut shape). Nanotori have many unique properties and among them magnetic moments 1000 times larger than previously expected for certain specific radii is one. Depending on the radius of the torus and the radius of the tube, properties such as magnetic moment, thermal stability, etc. vary widely.

Nanobud

Figure 2.8   A stable nanobud structure

Carbon nanobuds are a newly discovered material. It combines two previously discovered allotropes of carbon: carbon nanotubes and fullerenes. In this new material, fullerene-like "buds" are covalently bonded to the outer sidewalls of the underlying carbon nanotube (Figure 2.8). This being a hybrid material, it has useful prperties of both fullerenes and carbon nanotubes. They have been found to be exceptionally good field emitters.

Nanoflower

The first nanoflower was created in Japan. It was the accidental outcome of an experiment on nanotubes.

These are flower-shaped nanochemicals ."Flowers" are less than one-thousandth of the width of a human hair. Nanometre scale wires of a silicon–carbon material (silicon carbide) are grown from tiny droplets of a liquid metal (Gallium) on a silicon surface. When grown, the images are like flowers and hence named as nanoflowers.

PROPERTIES

Carbon nanotubes have certain special properties which are responsible for their utilization in various practical applications. The properties are:

1. Strength
2. Kinetic property
3. Electrical property
4. Thermal property

Strength

The strongest and stiffest materials on earth are carbon nanotubes, in terms of tensile strength and elastic modulus respectively. The formation of covalent bonds between the individual carbon atoms are responsible for the strength of the carbon nanotubes Multi-walled carbon nanotube can have a tensile strength of 63 gigapascals (GPa). (Ability to endure weight of 6300 kg on a cable with cross-section of 1 mm^2 is the strength.) The tubes can develop plastic deformation which is permanent because of the excessive tensile strain. This deformation begins at strains of approximately 5%. This can increase the maximum strain before fracture by releasing strain energy. CNTs are not that strong when under compression. When placed under compressive, torsional or bending stress, they tend to undergo buckling because of their hollow structure and high aspect ratio.

Kinetic Property

MWNTs or multiple concentric nanotubes precisely are nested within one another. They exhibit a striking telescoping property whereby an inner nanotube core may slide, almost without friction, within its outer nanotube shell. Thus it creates an atomically perfect linear or rotational bearing. This is one of the first practical examples of molecular nanotechnology wherein the precise positioning of atoms are used to create useful machines. The world's smallest rotational motor is created using this property. Gigahertz mechanical oscillator is also envisaged in future.

Electrical Property

The structure of a nanotube strongly affects its electrical properties because of the symmetry and unique electronic structure of graphene. A nanotube may be metallic or a semiconductor. In theory, metallic nanotubes can have an electrical current density more than 1,000 times greater than metals such as silver and copper.

Thermal Property

All nanotubes are very good thermal conductors along the tube, exhibiting a property known as "ballistic conduction". They are good insulators laterally to the tube axis. It is predicted that carbon nanotubes will be able to transmit up to 6000 Watts mpk at room temperature. Copper, a well-known metal for its good thermal conductivity, only transmits 385 $Wm^{-1} K^{-1}$. The temperature stability of carbon nanotubes can be up to 2800°C in vacuum and about 750°C in air.

DEFECTS OF NANOTUBES

When we are using any material, the existence of defects affects the material properties. Defects are possible in the form of atomic vacancies. High levels of such defects affect and lower the tensile strength by up to 85%. Stone Wales defect is another form of defect that may occur in carbon nanotubes. It creates a pentagon and heptagon pair by rearrangement of the bonds. Stone Wales defects cause phonon scattering over a wide range of frequencies, that in turn leads to a greater reduction in thermal conductivity. The tensile strength of the tube depends on its weakest segment due to the very small structure of CNTs which is comparable to a chain, where a defect in a single link lowers the strength of the entire chain.

Electrical properties of the tubes are affected due to defects and the result is the lowered conductivity through the defective region of the tube. Some defect formation in armchair-type tubes (which can conduct electricity) can cause the region surrounding that defect to become semiconducting. Single monoatomic vacancies can induce magnetic properties.

Thermal properties of the tubes are also heavily affected because of defects. Such defects lead to phonon scattering. The phonon scattering in turn increases the relaxation rate of the phonons and reduces the mean free path. The result is reduction of the thermal conductivity of nanotube structures. Phonon transport simulations indicate the substitutional defects, i.e., the substitution by nitrogen or boron and that will primarily lead to scattering of high frequency optical phonons.

One-Dimensional Transport

Electron transport in carbon nanotubes will take place through quantum effects due to their nanoscale dimensions. It will propagate only along the axis of the tube. This special transport property of carbon nanotubes is often referred to as "one-dimensional".

Toxicity

Important focus of CNTs' toxicity is related to the development of human cancer. So, nanotubes may be responsible for the development of cancer is basis of research studies. The toxicity of carbon nanotubes in relation to human beings is an important and most pressing question in nanotechnology. Various scientific tests on cells have confusing results and it is a spectrum of high toxicity to no toxicity. The reason for such contradictory results is that the materials used in various studies are not the same types of CNTs due to the available processes developing different types. Impurities such as cobalt and nickel which have proved toxic to human beings may be the true causes of the effects. The ranges of physical and chemical properties of CNTs (e.g. surface area, zeta potential) are not often controlled while doing toxicology studies. A study shows that once CNTs are inside the cell, they accumulate in the cytoplasm and cause cell death. Whether cell death is caused by CNTs or impurities is debatable. When carbon nanotubes are injected in the lungs of mice, the specialized cells in the lungs are incapable of disposing them because these tubes are too big for the cells to engulf. That leads to constitutive inflammation[5] (constant and active) which is a hallmark of precancerous symptom. High or extreme doses are required to achieve this response. The hydrophobicity of unfunctionalized carbon nanotubes causes agglomeration of tubes into larger bundles or particles. Because of their size these bundles themselves can cause suffocation of mice rather than toxic effects. The length of carbon nanotubes is the critical factor in toxicity or lack of biocompatibility along with the biological (and biochemical) environment. The environment may improve solubility in aqueous solutions due to protein adsorption to the carbon nanotube surface and this may be responsible for the toxicity. Buckypaper—which is a mat of carbon nanotubes compressed to a paper-like form, is useful for the growth of various cell types without having any toxic effects.

SYNTHESIS

Techniques useful to produce nanotubes in sizeable quantities are:

1. Arc discharge
2. Laser ablation
3. High pressure carbon monoxide (HiPCO)
4. Chemical vapour deposition (CVD)

The above processes take place in vacuum or with process gases. CVD growth of CNTs can take place in vacuum as well as at atmospheric pressure. Large quantities of nanotubes can be synthesized by the above methods. Advances in catalysis and continuous growth processes are making CNTs more commercially viable.

The catalysts or methods involved in forging Damascus steel (a forging technique lost to time) if studied, may provide vital hints for manufacturing nanotubes cheaply, since CNTs were recently discovered to be a component of that ancient sword metal.

Arc Discharge

During an arc discharge, by using a current of 100 A, that was intended to produce fullerenes, nanotubes were observed in the carbon soot of graphite electrodes in 1991. The carbon contained in the negative electrode sublimates due to the high temperatures caused by the discharge during this process. Arc discharge method has been the most widely used method of nanotube synthesis, because nanotubes were initially discovered using this technique.

The yield for this method is up to 30 per cent by weight. It produces both single- and multi-walled nanotubes with lengths of up to 50 μm.

Laser Ablation

A graphite target is vaporized by a pulsed laser in a high temperature reactor while an inert gas is bled into the chamber, in the laser ablation process. As the vaporized carbon condenses, the nanotubes develop on the cooler surfaces of the reactor. A water-cooled surface can be included in the system to collect the nanotubes.

This method has a yield of around 70%. It produces primarily single-walled carbon nanotubes with a controllable diameter determined by the reaction temperature. It is more expensive than either arc discharge or chemical vapour deposition.

High Pressure Carbon Monoxide (HiPCO)

The high-pressure carbon monoxide (HiPCO) process, developed at Rice University, has been reported to produce single-walled carbon nanotubes from gas-phase reactions of iron carbonyl in carbon monoxide at high pressures (10–100 atm). Computational modelling is used to develop an understanding of the HiPCO process.

The HiPCO process is a relatively clean process, i.e., SWNT is obtained without the graphitic deposits and amorphous carbon, and more importantly, the process has the potential for producing SWNT in large quantities since it is rom a gas-phase reaction (currently 10 g a day). In the HiPCO process, carbon monoxide mixed with a small amount of iron pentacarbonyl $\left(-Fe(CO_5^-)\right)$ is heated to produce SWNT. The products of thermal decomposition of Fe $(CO)_5$ react to produce iron clusters in gas phase. These metal clusters serve as nuclei upon which SWNT nucleate and grow.

Chemical Vapour Deposition (CVD)

During CVD, a substrate is prepared with a layer of metal catalyst particles, most commonly nickel, cobalt, iron, or a combination or by other ways, including reduction of oxides or oxide solutions. CVD is the most commonly used method for the commercial production of carbon nanotubes (Figure 2.9). During the growth process (plasma-enhanced chemical vapour deposition), a plasma is generated b the application of a strong electric field. The nanotube growth will follow the direction of the electric field in the plasma generated.

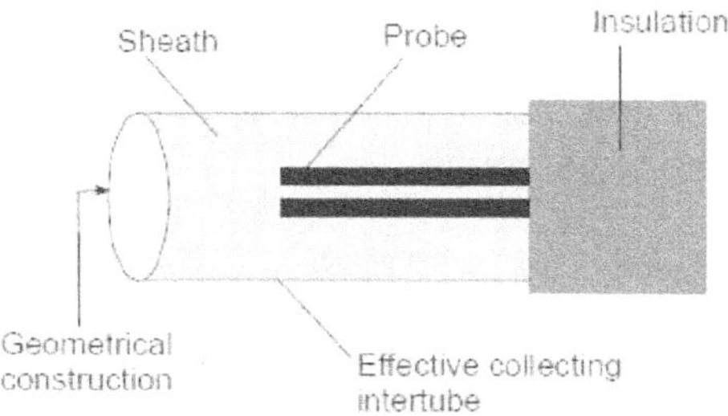

Figure 2.9 Instrument used for the plasma-enhanced chemical vapour deposition

Out of the various methods for nanotube synthesis, CVD is ideally suited for industrial scale deposition in terms of its price/unit ratio. There are positive points to the CVD synthesis of nanotubes. Compared to the other methods, CVD is capable of growing nanotubes directly on a desired substrate. But the nanotubes must be collected in the other growth techniques and then be used in desirable substrate. The growth sites in CVD can be controlled by careful deposition of the catalyst. Vertically aligned nanotubes can be produced only by CVD at the moment. A high-efficiency CVD technique for growing carbon nanotubes from camphor was developed in 2007.

NATURAL SOURCES

Fullerenes and carbon nanotubes are not only produced in high-tech laboratories but can also be commonly formed in common places of human activity like ordinary flames and by burning of methane, ethylene and benzene. They have also been found in soot from both indoor and outdoor air. Because of the highly uncontrolled environment in which they are produced, naturally occurring fullerenes are highly irregular in size and quality. They have got some applications. Due to lack of high degree of uniformity necessary to meet many needs of both research and industry, their use is limited. Recently efforts are being made for producing more uniform carbon nanotubes in controlled flame environments.

STRUCTURE

The strength and flexibility of carbon nanotubes can be used in controlling other nanoscale structures and they are possible tools in this utility. The highest tensile strength of an individual multi-walled carbon nanotube has been tested to be 63 GPa. Some carbon nanotubes are present in Damascus steel and the nanotubes possibly were responsible for the legendary strength of the (almost ancient) swords made of it. This was elaborated in the journal *Nature* in 2006.

India's mastery of nanotechnology even some three to four hundred years ago is revealed by recent studies. For the manufacture of Damascus sword in Turkey, which was extensively used as weapons in the battlefields till the 19th century, the raw material wootz steel was exported from India.

APPLICATIONS OF CARBON NANOTUBES

Carbon nanotubes have wide and potential applications. The most important applications are enumerated. This is a database for the researchers to pursue research in nanotubes.

Structural materials used in day-to-day life

- Waterproof tear-resistant cloth fibres are being produced for the manufacture of newer type of clothes.
- Combat jackets that use carbon nanotubes as ultra strong fibres.
- For the production of concrete with increased tensile strength and for avoiding crack propagation.
- Polyethylene with increased elastic modulus by 30%.
- Sports equipment like stronger and lighter tennis rackets, bike parts, golf balls, golf clubs, golf shaft and baseball bats.
- *Space elevator*   This will be possible only if tensile strengths of more than about 70 GPa can be achieved. Oxygen free radical in the Earth's upper atmosphere would erode carbon nanotubes at some altitudes. So, a space elevator constructed of nanotubes would need to be protected using protective coating. Carbon nanotubes in other applications need no such surface protection.
- *Ultrahigh-speed flywheels*   The high strength/weight ratio will enable very high speeds to be achieved.
- *Bridges*   In suspension bridges instead of steel.

Applications in electromagnetic field

❖ *Artificial muscles*   When a voltage is applied to electroactive polymers (EAPs), their shape is modified. They are useful as actuators or sensors. As actuators, they are able to undergo a large amount of deformation while sustaining large forces. They are often called artificial muscles, due to the similarities with biological tissues in terms of achievable stress and force. They have the potential for application in the field of robotics, where large linear movement is often needed.

❖ *Buckypaper*   This is made up of a thin sheet of nanotubes that are 250 times stronger than steel and 10 times lighter, and could be used as a heat sink for chipboards, a backlight for LCD screens or as a faraday cage to protect electrical devices/aeroplanes.

❖ *Chemical nanowires*   Carbon nanotubes can also be used to produce nanowires of other metals and chemicals, such as gold or zinc oxide. These nanowires are further used to cast nanotubes of other chemicals, such as gallium nitride. They have very different properties from CNTs, e.g. gallium nitride nanotubes are hydrophilic, while CNTs are hydrophobic.

❖ *Computer circuits*   A nanotube formed by joining nanotubes of two different diameters end to end can act as a diode. This gives the possibility of constructing electronic computer circuits entirely out of nanotubes. CNTs can also be used to dissipate the heat from tiny computer chips, due to their good thermal properties. The length of the longest electricity-conducting circuit is within millimetres.

❖ *Conductive films*   Mechanically more robust transparent, electrically conductive films of carbon nanotubes to replace indium tin oxide (ITO) in LCD touch screens, and photovoltaic devices were developed. They are ideal for high reliability touch screens and flexible displays. Printable water-based inks of carbon nanotubes are the next step to enable the production of these films to replace ITO. Nanotube films have got futuristic use in displays for computers, cell phones, PDAs, and ATMs.

❖ *Electric motor brushes*   Conductive carbon nanotubes are in use for several years in brushes for commercial electric motors. Instead of traditional carbon black, which is mostly impure, spherical carbon fullerenes, carbon nanotubes are used. The nanotubes improve electrical and thermal conductivity because they stretch through the plastic matrix of the brush. This allows the carbon filler to be reduced from 30% down to 3.6%, and so more matrix is present in the brush. Nanotube composite motor brushes are better lubricated (from the matrix), cooler-running (both from better lubrication and superior thermal conductivity), less brittle (more matrix, and fibre reinforcement), stronger and more accurately mouldable (more matrix). Brushes are a critical failure points in electric motors. When CNT brushes are used, the failure rate is negligible. So, electric motor brushes become economical out of all applications of CNT.

◈ *Light bulb filament* It is used as an alternative to tungsten filaments in incandescent lamps.

◈ *Magnets* MWNTs coated with magnetite are useful as magnets.

◈ *Optical ignition* A layer of 29% iron-enriched SWNT is placed on top of a layer of explosive material such as PETN[6] (Pentaerytritol tetra nitrate) and can be ignited with a regular camera flash.

◈ *Solar cells* Carbon nanotube diode (Figure 2.10) has a photovoltaic effect. Nanotubes can replac ITO in some solar cells. They act as a transparent conductive film in solar cells to allow light to pass to the active layers and generate photocurrent.

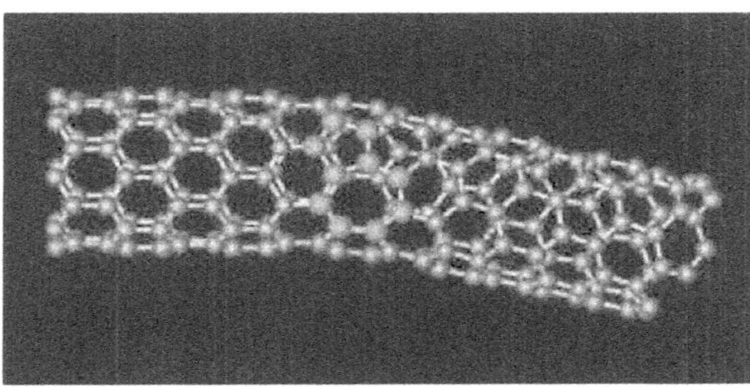

Figure 2.10 Two carbon nanotubes joined to form a diode

◈ *Superconductor* Nanotubes have been shown to be superconducting at low temperatures which can be used for various applications.

◈ *Ultracapacitors* To dramatically increase the surface area and therefore energy storage ability, nanotubes are bound to the charge plates of capacitors.

◈ *Displays* Extremely fine electron guns, used as miniature cathode ray tubes in thin, high-brightness, low-energy, low-weight displays are developed. This display consists of a group of many tiny CRTs, each providing the electrons to hit the phosphor of one pixel, unlike one giant CRT whose electrons are aimed using electric and magnetic fields as of now. They are called as field emission displays (FEDs).

◈ *Transistor* Developed by various researchers.

Applications in chemical field

◈ *Air pollution filter* Filtering carbon dioxide from power plant emissions is a future application of nanotube membranes.

◈ *Biotech container* Nanotubes can be used as a biotech container making the possibility of applications in biotechnology.

◈ *Hydrogen storage* They have the potential to store between 4.2 and 65% hydrogen by weight and so potential use of carbon nanotubes for hydrogen storage is of futuristic application.

◈ *Water filter* Nanotube membranes for use in filtration have been developed. The tubes are so thin that small particles (like water molecules) can pass through them, while larger particles (such as the chloride ions in salt) are blocked. This technique can purportedly reduce desalination costs by 75%.

Mechanical applications

◈ *Oscillator* At the present juncture, fastest known oscillators (>50 GHz) are nanotubes.

◈ *Nanotube membrane* Liquid flows up to five orders of magnitude faster than predicted by classical fluid dynamics in nanotube membranes and so useful in various applications.

◈ *Slick surface* Slicker than teflon and waterproof are the materials developed out of nanotubes and useful as materials of day-to-day life.

Applications in electrical circuits

Carbon nanotubes have unique dimensions to an unusual current conduction mechanism and that make them ideal components of electrical circuits. As of now, there is no accepted way to arrange carbon nanotubes into a circuit. Even if precisely positioned, there remains the problem of controlling the types of nanotubes namely, metallic, semiconducting, single-walled and multi-walled while being produced for better utilization.

Metallic and semiconducting nanotubes CNTs of multi-walled structures with outermost shell diameters exceeding 10 nm can be used as light-emitting semiconductors.

Carbon nanotube interconnects Very-large-scale integration (VLSI) interconnects of the future are metallic CNTs because of their desirable properties of high thermal stability, high thermal conductivity and large current carrying capacity.

Carbon nanotube transistors Field effect transistors (CNTFETs) are fabricated using semiconducting CNTs, which show promise due to their superior electrical characteristics over silicon-based MOSFETs. It is superior in terms of sub-threshold slope which is a very important property for low-power applications.

Challenges in electronic design and design automation CNT devices and interconnects have been separately shown to be promising in their own respects. But there are few efforts to successfully combine them in a realistic circuit for future applications.

As fibre and film High tensile strength fibres is an important research area of application.

The Cambridge-MIT Institute developed a method of creating carbon nanotube fibre continuously at the speed of several centimetres per second just as nanotubes are produced. One thread of carbon nanotubes was more than 100 m long. The resulting fibres are electrically conductive and as strong as ordinary textile threads.

Drug delivery vessel The nanotube's versatile structure is useful for a variety of tasks in and around the human body. The carbon nanotube is often used as a vessel for transporting drugs into the body apart from its uses in cancer cases. The nanotube usage can allow lower drug dosage by localizing its distribution, which will significantly cut costs to pharmaceutical companies and their consumers. The nanotube commonly carries the drug in one of two methods:

1. The drug can be attached to the side of the nanotube or can be attached in the end to trail behind the nanotube.

2. The drug can be placed inside the nanotube for delivery of the drugs.

Both of these methods are effective and useful for the delivery and distribution of drugs inside the body for effective therapy.

CURRENT APPLICATIONS

◇ Presently, carbon nanotubes are used as bulk nanotubes, a mass of rather unorganized fragments of nanotubes. Bulk nanotube materials cannot achieve a tensile strength similar to that of individual tubes. But such composites have got strengths sufficient enough for many applications. Bulk carbon nanotubes have already been used as composite fibres in polymers that improves the mechanical, thermal and electrical properties of the bulk product. Carbon nanotubes have been successfully used in the construction of handlebars for mountain bikes.

◇ The solar cell has been developed using a carbon nanotube complex, formed by carbon nanotubes and combines them with tiny carbon buckyballs to form snakelike structures. Buckyballs trap electrons, but cannot make electrons flow. Sunlight is used to excite the polymers and the buckyballs will grab the electrons. Nanotubes, like that of copper wires, will then be able to make the electrons or current to flow.

◇ Carbon nanotubes have been implemented in nanoelectromechanical systems (NEMS) which includes mechanical memory elements (NRAM) and nanoscale electric motors.

◇ Carbon nanotubes have been projected as a possible gene delivery vehicle. It is used in combination with radio frequency fields to destroy cancer cells.

◈ Hydrogen sensor is an electronic device that was made by integrated carbon nanotubes on a silicon platform. Sensor applications such as in the field of carbon dioxide, nitrous oxide, glucose, DNA detection, etc. are being developed for various applications.

◈ A nanoradio, a radio receiver consisting of a single nanotube, was created in 2007. A sheet of nanotubes operating as a loudspeaker if an alternating current is applied was demostrated in 2008. The sound is produced thermoacoustically and not by vibration.

◈ Carbon nanotubes are possibly having the strength of diamond. Research is on to weave them into clothes to create stab-proof and bulletproof clothing. The nanotubes would effectively stop the bullet from penetrating the body.

◈ A flywheel made of carbon nanotubes could be spun at extremely high velocity on a floating magnetic axis. That can potentially store energy at a density approaching that of conventional fossil fuels. Energy can be added to and removed from flywheels very efficiently in the form of electricity. This might offer a way of storing electricity, making more efficient electrical grid and variable power suppliers (like wind turbines). They will be more useful in meeting where the energy needs are more. The practicality of this application depends heavily upon the cost of making massive, unbroken nanotube structures and also their failure rate under stress.

COLLOIDAL GOLD

INTRODUCTION

Figure 2.11 Aqueous colloidal gold

Colloidal gold, otherwise known as "nanogold", is a suspension (or colloid) of sub-micrometre-sized particles of gold in a fluid, usually water (Figure 2.11). The liquid is usually observed as either an intense red colour (for particles less than 100 nm), or a dirty yellowish olour (for larger particles). The nanoparticles themselves can come in a variety of shapes like spheres, rods, cubes and caps and are the more frequently observed ones.

HISTORY

In ancient times, Elixir of Life, a potion made from gold, was prepared. Whether it was actually manufactured or not is possible to answer at the present juncture. In Tamil Nadu, it was described in Siddha Medicine in ancient times. Colloidal gold has been used since ancient Roman times to colour glass in intense shades of yellow, red, or mauve, depending on the concentration of gold. The alchemist Paracelsus claimed to have created a potion called Aurum Potabile (Latin *potable gold*) in the 16th century. The glass-colouring process was refined by Andreus Cassius and Johann Kunchel in the 17ᵗʰ century. John Herschel invented a photographic process called Chrysotype (from the Greek word for gold), that used colloidal gold to record images on paper in 1842. Paracelsus' work is known to have inspired Michael Faraday to prepare the first pure sample of colloidal gold and was called activated gold, in 1857. He used phosphorus to reduce a solution of gold chloride. The colour was due to the minute size of the gold particles being recognized for the first time by Faraday.

SYNTHESIS

The synthesis of colloidal gold was originally used as a method of staining glass since ancient times. Michael Faraday's work (1850) was the starting point for the modern scientific evaluation of colloidal gold. Colloidal gold is being researched, with applications in a wide variety of areas, including electronics, nanotechnology and the synthesis of novel materials with unique properties. This is due to the unique optical, electronic, and molecular recognition properties of gold nanoparticles.

Liquid Chemical Methods

Usually, gold nanoparticles are produced in a liquid by reduction of chloroauric acid ($H[AuCl_4]$) called "liquid chemical methods". More advanced and precise methods also exist. $H[AuCl_4]$ is dissolved and the solution is rapidly stirred while a reducing agent is added. This makes Au^{3+} ions to be reduced to neutral gold atoms. With more formation of gold atoms, the solution becomes supersaturated. So, gold gradually starts precipitating in the form of sub-nanometre particles. Subsequently, the rest of the gold atoms that form stick to the existing particles. If the solution is stirred vigorously well, the particles will be fairly uniform in size.

For the prevention of the particles from aggregating, a stabilizing agent that sticks to the nanoparticle surface is usually added. They can be functionalized by combining with various organic ligands to create organic–inorganic hybrids with advanced functionality.

Turkevich *et al.* Method

It was created by J. Turkevich *et al.* in 1951. It was refined by G. Frens in 1970s. This method is the simplest one available at present. The evolution of the spherical gold nanoparticles in the Turkevich reaction has been elucidated recently. Extensive networks of gold nanowires are formed as a transient intermediate. They are responsible for the dark appearance of the reaction solution before it turns ruby-red.

Brust *et al.* Method

This method was discovered by Brust and Schiffrin in the early 1990s. It is used to produce gold nanoparticles in organic liquids which are normally not miscible with water (like toluene). The gold nanoparticles will be around 5–6 nm in this method of production.

Sonolysis

This is the method for the experimental generation of gold particles. This process is based on ultrasound, the reaction of an aqueous solution of $H[AuCl_4]$ with glucose and the morphology obtained is that of nanoribbons with width 30–50 nm and length of several micrometres.

HEALTH/MEDICAL APPLICATIONS

Colloidal gold has been successfully used as a therapy for arthritis. This was used in Siddha System of Medicine, an ancient system practised in Tamil Nadu and is followed even now in rheumatoid arthritis. The implantation of gold beads near arthritic hip joints in dogs has been found to relieve pain in a study.

The combination of microwave radiation and colloidal gold can destroy the beta-amyloid fibrils and plaque which are associated with Alzheimer's disease and was demonstrated in an *in vitro* experiment. The possibilities for numerous similar radiative applications are currently under research.

Gold nanoparticles are being investigated as carriers for drugs such as Paclitaxel, a drug used in cancer therapy. The administration of hydrophobic drugs requires encapsulation. It is found that nanosized particles are particularly efficient in evading the reticuloendothelial system. So, the delivery of the drug in place is certain and adequate especially in cancer patients.

In cancer research, colloidal gold can be used to target tumours. They can be detected using SERS (Surface Enhanced Raman Spectroscopy) *in vivo*. This helps to make effective dosage for better therapy with least side effects.

Nanogold-coated bacteria can be used for electronic wiring (2005). The reduction of hydrogen tetrachloroaurate by sodium borohydride in the presence of one of the enantiomers of penicillamine results in optically active colloidal gold particles.

NANOSTRUCTURE

INTRODUCTION

A nanostructure is an object of intermediate size between molecular and microscopic (micrometre-sized) structures. Obviously, they have got the dimensions under nanoscale. While nanostructures are described, it is essential to differentiate between the numbers of dimensions on the nanoscale.

1. Nanotextured surfaces have one dimension on the nanoscale, i.e., only the thickness of the surface of an object is between 0.1 and 100 nm.

2. Nanotubes have two dimensions on the nanoscale, i.e., the diameter of the tube is between 0.1 and 100 nm; its length could be much greater.

3. Spherical nanoparticles have three dimensions on the nanoscale, i.e., the particle is between 0.1 and 100 nm in each spatial dimension. The terms nanoparticles and ultrafine particles (UFP) often are used synonymously although UFP can reach into the micrometre range.

LIST OF NANOSTRUCTURES

These are the list of nanostructures which are presently described, used or researched. This will give some insight into the field of nanotechnology.

- Nanocages
- Nanocomposites
- Nanofabrics
- Nanofibre
- Nanoflower
- Nanofoam
- Nanomesh

- ❖ Nanopillar
- ❖ Nanopin film
- ❖ Nanoring
- ❖ Nanorod
- ❖ Nanoshell
- ❖ Sculptured thin film

NANOSCALE IRON PARTICLES

INTRODUCTION

The earth is covered by environmental contaminants. These contaminants include polychlorinated biphenyls (PCBs), chlorinated organic solvents and organochlorine pesticides. The present clean-up methods for these contaminants are costly and limited. Research is on in the field of environmental science finding less costly alternatives. Using nanoscale iron particles, for the cleaning process is an alternative method that has proven to be successful in laboratory research as well as in field test studies.

STRUCTURE

Nanoscale iron particles are small in size, usually 1–100 nm. Like most of the nanoscale materials, their large surface areas make them highly reactive, and also they are easily transportable through groundwater. This helps in *in situ* treatment. The nanoparticle-water slurry can be sprayed over the contaminated area and stay there for long periods of time. When exposed to oxygen and water, iron oxidizes.

RESEARCH

Although metallic iron nanoparticles remediate contaminants well, they tend to agglomerate on the soil surfaces. Carbon nanoparticles and water-soluble polyelectrolytes are used as supports to the metallic iron nanoparticles. The hydrophobic contaminants adsorb to these support materials and thus improve permeability in sand and soil. Apart from laboratory research, field studies have been conducted to establish the fact. The field test showed better water quality. This procedure also helps to control total chlorinated volatile organic compounds (VOCs) to avoid toxicity out of them.

SELF-ASSEMBLED MONOLAYERS

Self-assembled monolayers (SAMs) are surfaces that consist of a single layer of molecules on a substrate. Chemical vapour deposition and molecular beam epitaxy[7] are the techniques which are used to add molecules to a surface (often with poor control over the thickness of the molecular layer). Self-assembled monolayers can be prepared simply by adding a solution of the desired molecule onto the substrate surface and washing off the excess. A common example is an alkane thiol on gold. Sulphur has particular affinity for gold. More ordering of the assembly can take place over days or months, depending on the molecules involved.

A variety of other self-assembled monolayers can be formed, although there is always debate about the degree to which systems self-assemble. Alkyl thiols are known to assemble on many metals, including silver, copper, palladium, and platinum. Alkyl silane molecules (e.g. octadecyltrichlorosilane) are another well-known example of self-assembly on silicon oxide surfaces and are potentially of greater technical relevance than alkyl thiol assembly on metals. Alkyl carboxylates are known to assemble on a variety of surfaces, such as aluminium and mica. Silicon has been used through the reaction of silicon hydride surface and a radical generator, such as heat, UV or radical initiator molecule.

Applications

SAMs have several applications in scientific research; they tend to have quite different chemical kinetics than the same molecules in another form, because of their exposed, 2-dimensional distribution and as such are useful for some chemical and biochemical experiments. They can also be used for simulation of biological membranes and as substrates for cell culture. As technology develops to control the functional groups present inSAMs, either by direct deposition of molecules with those groups or by chemical modification of the layer, many other applications are also developing, for exampl, in nanoscale fabrication of electronics.

ENDNOTES

1. Corannulene is considered as a fragment of buckminsterfullerene.

2. Julian Voss-Andreae's creation of a 30' (9 m) diameter Buckyball sculpture (see Picture).

3. Sintering is a method for making objects from powder, by heating the material (below its melting point—solid state sintering) until its particles adhere to each other. Sintering is traditionally used for manufacturing ceramic objects, and has also found uses in such fields as powder metallurgy.

4. Arcology, from the words "architecture" and "ecology" is a set of architectural design principles aimed toward the design of enormous habitats (hyper structures) of extremely high human population density. These largely hypothetical structures, called "arcologies," would contain a variety of residential and commercial facilities and minimize individual human environmental mpact. They are often portrayed as self-contained or economically self-sufficient. Picture shows the Try2004 Hyper structure or Megacity as featured on the Discovery Channel's *Extreme Engineering* programs.

5. Constitutive inflammation   In cell biology, a constitutively active protein is a protein whose activity is constant and active.

6. PETN   This compound finds both medicinal use as a coronary vasodilator (often mixed with lactose to reduce sensitivity) and extensive use in the explosive industry, as a very high power booster in blasting caps and also as the filler indetonating fuse.

7. Molecular beam epitaxy (MBE) is one of several methods of depositing single crystals. It was invented in the late 1960s at Bell Telephone Laboratories by J. R. Arthur and Alfred Y. Cho.

REVIEW QUESTIONS

1. Describe in detail about nanomaterials.
2. Give a detailed account of fullerenes.
3. Discuss the properties of fullerenes.
4. Describe nanoparticles in detail.
5. Discuss the classification of nanoparticles.
6. Explain the properties of nanoparticles.
7. How are nanoparticles produced?
8. Discuss in detail about carbon nanotubes.
9. Discuss in detail about synthesis of carbon nanotubes.
10. Discuss in detail about applications of carbon nanotubes.
11. Write short notes on:
    i. Types of carbon nanotubes
    ii. Single-walled nanotubes
    iii. Multi-walled nanotubes
    iv. Fullerite
    v. Properties of carbon nanotubes
    vi. Nanochemistry
    vii. Classification of nanomaterials
    viii. Variations of fullerenes
    ix. Buckyballs
    x. Nanotubes
    xi. Characterization of nanoparticles
12. Discuss the toxicity of carbon nanotubes.
13. What is chemical vapour deposition?
14. Give the applications of CNT in
    i. electromagnetic field

    ii.   chemical field

    iii. mechanical field

    iv. eectrical circuits

15. Discuss in detail about colloidal gold.

16. Write short notes on:

    i.   Colloid gold

    ii.   Nanostructures

    iii. Nanoscale iron particles

    iv. Remediation

    v.   Self-assembled monolayers

REFERENCES

1. *"Carbon Nanotubes from Camphor*: An Environment Friendly Nanotechnology" (PDF). 2007.p. 643.

2. Appenzeller, J. *et al.* (2005) "Comparing Carbon Nanotube Transistors - The Ideal Choice: A Novel Tunneling Device Design," IEEE TED, Vol. 52, No. 12, pp. 2568–2576.

3. Belluci, S. (2005). "Carbon nanotubes: Physics and applications." *Phys. Stat. Sol.* (c) 2(1):34–47.

4. Buzea, C. (2007). "Nanomaterials and nanoparticles: Sources and toxicity." *Biointerphases.* 2: MR17.

5. Collins, G. Philip and Phaedon Avouris. (2000). "Nanotubes for electronics." *Scientific American,* 67, 68, and 69.

6. Cristina Buzea, Ivan Pacheco, and Kevin Robbie. (2007). "Nanomaterials and nanoparticles: Sources and toxicity." *Biointerphases.* 2: MR17–MR71.

7. Dekker, Cees. (1999). "Carbon nanotubes as molecular quantum wires." *Physics Today.* 52 (5):22–28.

8. Ebbesen, T.W. and Ajayan, P.M. (1992). "Large-scale synthesis of carbon nanotubes." *Nature.* 358: 220–222.

9. Fahlman, B.D. (2007). *Materials Chemistry.* Vol. 1. Springer, Mount Pleasant, MI. pp 282–283.

10. Gary, L. Miessler. and Donald, A. Tarr. (2004). *Inorganic Chemistry,* 3rd edn. Pearson Education International.

11. Huhtala Maria, Kuronen Antti and Kaski Kimmo. (2002). "Carbon nanotube structures: molecular dynamics simulation at realistic limit." *Computer Physics Communications.* 146: 30.

12. Iijima, Sumio. (1991). "Helical microtubules of graphitic carbon." *Nature.* 354: 56–58.

13. Inami, N. *et al.* (2007). "Synthesis-condition dependence of carbon nanotube growth by alcohol catalytic chemical vapor deposition method." *Sci. Technol. Adv. Mater.* 292.

14. Kolosnjaj, J., Szwarc, H. and Moussa, F. (2007). "Toxicity studies of carbon nanotubes." *Adv. Exp. Med. Biol.* 620: 181–204.

15. Krätschmer, W. *et al.* (1990). "Solid C60: a new form of carbon." *Nature.* 347: 354.

16. Kreupl, F. *et al.* (2002). "Carbon nanotubes in interconnect applications." *Microelectronic Engineering.* 64: 399–408.

17. Kroto, H.W., Heath, J.R., O'Brien, S.C., Curl, R.F. and Smalley, R.E. (1985). "C_{60}: Buckminsterfullerene." *Nature.* 318: 162–163.

18. Meo, S.B. and Andrews, R. (2001). "Carbon nanotubes: Synthesis, properties, and applications." *Crit. Rev. Solid State Mater. Sci.* 26(3):145–249.

19. Mori, T. *et al.* (2006). "Preclinical studies on safety of fullerene upon acute oral administration and evaluation for no mutagenesis." *Toxicology.* 225: 48–54.

20. Sanderson, K. (2006). "Sharpest cut from nanotube sword." *Nature.* 444: 286.

21. Wang, N. *et al.* (2000). "The smallest carbon nanotube." *Nature.* 408: 50.

22. Ying, Jackie. (2001). *Nanostructured Materials.* Academic Press, New York.

23. Zhang, Wei-xian. (2003). "Nanoscale iron particles for environmental remediation: An overview." *Journal of Nanoparticle Research.* 5: 323–332.

24. Zheng, L.X. *et al.* (2004). "Ultralong single-wall carbon nanotubes." *Nature Materials.* 3: 673–676.

Chapter 3
Nanophysics

Nanophysics is a branch of nanotechnology which deals with the study of the physical properties of objects in the nanoscale size range. Some interesting physical properties studied under nanophysics include conductivity of small numbers of atoms and molecules, the forces arising between nanoscale objects, and the transition between the quantum behaviour exhibited by a few atoms and the bulk properties of a large number of atoms. Scanning tunneling and scanning force microscopes are the instruments that allow the study of nanometre-scale objects.

The following studies are also dealt under nanophysics work.

1. Behaviour of quantum particles on nanoscale.

2. Nanoscale semiconductor devices

3. Quantum information processing

4. Quantum physics to explore innovative applications in quantum information processing

5. Quantum wires and low-dimensional electronic systems

6. Materials confinement on nanoscale

7. Quantum dots and coupled quantum dots, which are exciting artificial atoms and molecules developed by physicists, materials scientists and chemists over the past decade.

This chapter deals with quantum dots in detail, quantum confinement in semiconductors, quantum wire, quantum well, quantum point contact, nanocrystal solar cell and nanocrystals.

QUANTUM DOT

Quantum dots, otherwise a form of nanocrystals, are a non-traditional type of semiconductors with limitless applications in many industries. A quantum dot is defined as a semiconductor nanostructure (Figure 3.1). It has got the capacity to confine the motion of conduction band electrons[1], valence band holes[2], or excitons[3] (bound pairs of conduction band electrons and valence band holes) in all three spatial directions. The confinement can be due to

1.   electrostatic potentials which are generated by external electrodes, doping, strain or impurities,

2.   the presence of an interface between different semiconductor materials (e.g. in core-shell nanocrystal systems),

3.   the presence of the semiconductor surface (e.g. semiconductor nanocrystal) or

4.   a combinaton of these

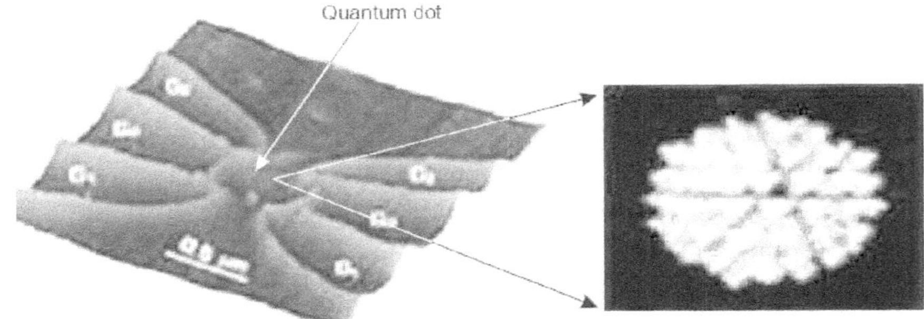

Figure 3.1   Quantum dot

A quantum dot possesses a discrete quantized energy spectrum. The corresponding wave functions are spatially localized within the quantum dot, but extend over many periods of the crystal lattice. A quantum dot contains a small finite number (of the order of 1–100) of conduction band electrons, valence band holes, or excitons, i.e., a finite number of elementary electric charges. Researchers have studied quantum dots in transistors, solar cells, light-emitting diodes (LEDs), diode lasers, as agents for medical imaging and hope to use them as qubits.[4]

DESCRIPTION

Small quantum dots like colloidal semiconductor nanocrystals are as small as 2 to 10 nm. That corresponds to 10 to 50 atoms in diameter and also a total of 100 to 100,000 atoms within the quantum dot volume. Self-assembled quantum dots are typically between 10 and 50 nm in size. To describe the size and structure of quantum dots, that are 10 nm in diameter, nearly 3 million quantum dots could be lined up end to end and fit within the width of a human thumb.

QUANTUM CONFINEMENT IN SEMICONDUCTORS

When a crystal is having the same or smaller exciton Bohr radius[5] of its constituent compound, then it is in a state of quantum confinement. Quantum confinement is the state of a material with a set of conditions. Under quantum confinement, energy levels are discrete in nature. By definition, quantum dots are in a state of quantum confinement.

In an unconfined state (bulk), semiconductors usually have an electron-hole pair which is typically bound within the exciton Bohr radius. If the electron and hole are constrained further or reduced in distance between them, then the semiconductor's properties change. This effect is called as quantum confinement. This is a key feature in many emerging electronic structures.

Quantum dots can be compared to other semiconductor nanostructures:

1.   Quantum wires would confine the motion of electrons or holes in two spatial directions. They allow free propagation in the third direction.

2.   Quantum wells would confine the motion of electrons or holes in one direction. They allow free propagation in two directions.

Quantum dots containing electrons are like atoms: both have a discrete energy spectrum and bind a small number of electrons. In contrast to atoms, the confinement potential in quantum dots is not always in spherical symmetry. The confined electrons are not moving in free space, but in the semiconductor host crystal only. The quantum dot host material's properties especially its band structure, play an important role for all quantum dot properties. If compared, energy scales of atoms are of the order of ten electron volts whereas that of quantum dots, only one millielectron volt. Quantum dots with a nearly spherical symmetry or flat quantum dots with nearly cylindrical symmetry can show shell filling according to the equivalent of Hund's rules for atoms[6]. They are also called as "artificial atoms".

OPTICAL PROPERTIES

The energy levels of small quantum dots can be probed by optical spectroscopy techniques like that of atoms. In quantum dots that confine electrons and holes, the interband absorption edge is blue-shifted[7] due to the confinement compared to the bulk material of the host semiconductor material. So, quantum dots of the same material, but with different sizes, can emit light of different colours depending on the size.

Quantum dots are very much significant for optical applications due to high quantum yield.

One of the optical features of small excitonic quantum dots is colouration easily visible to the unaided eye. Quantum dot's intrinsic energy signature is due to the material making it, especially the size which is correlated to colouration. The larger the dot is, more red is the fluorescence. The smaller the dot is, more blue is the fluorescence. The colouration is directly related to the energy levels of the quantum dot. The band gap energy determines the energy (and hence colour) of the fluoresced light. It is inversely proportional to the square of the size of the quantum dot. Larger quantum dots have more energy levels and are more closely spaced. This allows the quantum dot to absorb photons containing less energy and nearer to the red end of the spectrum. The lifetime of fluorescence is determined by the size. Larger quantum dots with more closely spaced energy levels can trap the electron-hole pair. Those electron-hole pairs in larger dots live longer and so, these larger dots show a longer lifetime.

The changeability in the size of quantum dots is advantageous for many applications. Larger quantum dots having spectra shifted towards the red compared to smaller dots, exhibit less pronounced quantum properties. On the other hand, the smaller particles have advantage of quantum properties.

FABRICATION

1. Self-assembled quantum dots are typically between 10 and 50 nm in size. Quantum dots are defined by lithographically patterned gate electrodes, or by etching on two-dimensional electron gases in semiconductor heterostructures having lateral dimensions exceeding 100 nm.

2. Some quantum dots are having small regions of one material buried in another with a larger band gap. These are called core-shell structures, e.g. with CdSe in the core and ZnS in the shell or from special forms of silica called ormosil.

3. Sometimes quantum dots occur in quantum well structures spontaneously due to monolayer fluctuations in the well's thickness.

4. Self-assembled quantum dots spontaneously nucleate under certain conditions during molecular beam epitaxy (MBE) and metallo-organic vapour phase epitaxy (MOVPE). This occurs when a material is grown on a substrate to which it is not lattice-matched. The resulting strain produces coherently strained islands on top of a two-dimensional "wetting-layer". This growth mode is known as Stranski–Krastanov growth[8]. The islands can be subsequently buried to form the quantum dot. This fabrication method has got the potential for applications in quantum cryptography (i.e., single photon sources) and quantum computation. The cost of fabrication and the lack of control over positioning of individual dots are the main limitations of this method.

5. Individual quantum dots can be derived from two-dimensional electron or hole gases. They are present in remotely doped quantum wells or semiconductor heterostructures.

A thin layer of resist is coated on the sample surface. By lithography, lateral pattern is then defined in the resist. This pattern can then be transferred to the electron or hole gas. This is done by etching, or by depositing metal electrodes (lift-off process) that allow the application of external voltages between the electron gas and the electrodes. Quantum dots out of this fabrication method are mainly of use in experiments and applications involving electron or hole transport, i.e., an electrical current.

6. By controlling the geometrical size, shape, and the strength of the confinement potential, the energy spectrum of a quantum dot can be engineered. It is relatively easy to connect quantum dots by tunnel barriers to conducting leads. That allows the application of the techniques of tunnelling spectroscopy for their investigation. This is in contrast to atoms.

7. From electrostatic potentials, confinement in quantum dots can arise like the ones generated by external electrodes, doping, strain, or impurities.

MASS PRODUCTION

In large numbers, quantum dots are synthesized by means of a colloidal synthesis. Colloidal synthesis is by far the cheapest and the least toxic of all the different forms of synthesis and has the advantage of being able to occur at bench-top conditions.

Self-assembly of highly ordered arrays of quantum dots is done by electrochemical techniques. At an electrolyte–metal interface, a template is created by causing an ionic reaction. That results in the spontaneous assembly of nanostructures, including quantum dots, on the metal which is used as a mask for mesa-etching these nanostructures on a chosen substrate.

Pyrolytic synthesis is the other method, which produces large numbers of quantum dots that self-assemble into preferential crystal sizes.

APPLICATIONS

Due to their theoretically high quantum yield, quantum dots are much useful for optical applications. In electronic applications, they operate like a single-electron transistor and exhibit the Coulomb blockade effect.[9] Quantum dots are also useful as implementations of qubits for quantum information processing.

The ability to tune the size of quantum dots is an advantageous property for many applications. Larger quantum dots have spectra towards the red and exhibit less pronounced quantum properties. The smaller particles have spectra towards th blue and exhibit more pronounced quantum properties.

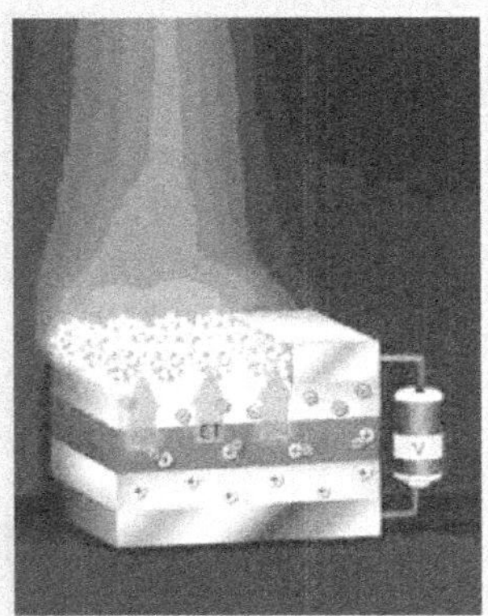

Wireless device

Researchers at Los Alamos National Laboratory have developed a wireless device that efficiently produces visible light, through energy transfer from thin layers of quantum wells to crystals above the layers.

The zero-dimensional aspect gives the quantum dots a sharper density of states compared to higher-dimensional structures. Because of that, they have superior transport and optical properties. So, they are investigated for use in diode lasers, amplifiers, and biological sensors.

Quantum dots can be excited within the locally enhanced electromagnetic field produced by the gold nanoparticles. This is observed from the surface plasmon resonance in the photoluminescence excitation spectrum of (CdSe) ZnS nanocrystals. High-quality quantum dots are very much useful for optical encoding and multiplexing applications. This is possible because of their broad excitation profiles and narrow/ symmetric emission spectra. The new generations of quantum dots have the ability for the study of intracellular processes at the single-molecule level, high-resolution cellular imaging, long-term *in vivo* observation of cell trafficking, tumour targeting, and diagnostics.

Computing Field

Quantum dot technology is of use in solid-state quantum computation. By applying small voltages to the leads, the flow of electrons through the quantum dot can be controlled. That helps to make precise measurements of the spin and other properties therein. Using several entangled quantum dots or qubits along with development of ways of performing operations using them, quantum calculations and the computers to do such calculations can be a reality in future.

Biology

In modern biological analysis, the role of various kinds of organic dyes is inevitable. Advances need more flexibility in the properties of dyes and the traditional dyes are not able to cope up with the necessity. Quantum dots have quickly filled this gap. They are found to be superior to traditional organic dyes on several counts. One of the most important is brightness (owing to the high-quantum yield) as well as their stability (much less photo-destruction). The irregular blinking of quantum dots is a minor drawback related to single-particle tracking. Tuning of the toxicity is the recent research area.

Cellular Imaging

Quantum dots are used for highly sensitive cellular imaging. The improved photostability of quantum dots is very useful for the acquisition of many consecutive focal-plane images that can be reconstructed into a high-resolution three-dimensional image. The real-time tracking of molecules and cells over extended periods of time is another application that takes advantage of the extraordinary photostability of quantum dot probes. Even after four months, researchers were able to observe quantum dots in lymph nodes of mice.

Semiconductor quantum dots are useful for *in vitro* imaging of pre-labelled cells. The ability to image single-cell migration in real time is expected to be important to several research areas such as embryogenesis, cancer metastasis, stem-cell therapeutics, and lymphocyte immunology. Quantum dots are useful for delivering a gene-silencing tool, known as siRNA, into cells, better than existing methods.

Quantum Dots for Tumour Targeting

Quantum dots for tumour targeting under *in vivo* conditions are attempted. There are two basic targeting schemes.

1. **Active targeting**   In this, quantum dots are made with tumour-specific binding sites. In that way, quantum dots are specifically bound to tumour cells only by the binding sites.
2. **Passive targeting**   This utilizes enhanced permeation and retention of tumour cells for the delivery of quantum dot probes.

Fast growing tumour cells have greater permeable membranes than healthy cells. These allow the leakage of small nanoparticles into the cell body. Since tumour cells lack an effective lymphatic drainage system, nanoparticle accumulation occurs.

Quantum dots as potential artificial fluorophore for intra-operative detection of tumours using fluorescence spectroscopy is also being researched which is one of the cutting-edge applications.

In vivo Toxicity

Quantum dot probes and their *in vivo* toxicity is an important point of concern. CdSe nanocrystals are highly toxic to cultured cells under UV illumination. The energy of UV irradiation is close to the covalent chemical bond energy of CdSe nanocrystals. Due to this, semiconductor particles can be dissolved, in a process known as photolysis, to release toxic cadmium ions into the culture medium. In the absence of UV irradiation, they have been found to be non-toxic if quantum dots are with a stable polymer coating. The excretion process of polymer-protected quantum dots from living organisms is not well explained. Before quantum dot applications in tumour or vascular imaging can be approved for human clinical use, these and related questions must be carefully examined.

Photovoltaic Devices

Quantum dots can increase the efficiency and reduce the cost of silicon photovoltaic cells. Quantum dots of lead selenide can produce seven excitons from one high-energy photon of sunlight (7.8 times the band gap energy) compared to photovoltaic cells with one exciton per high-energy photon, with high kinetic energy carriers losing their energy as heat. This would boost the maximum theoretical efficiency from 31% to 42%. Quantum dot photovoltaics would be cheaper to manufacture, as they are made "using simple chemical reactions."

Light-emitting Devices (LED)

Using quantum dots as light-emitting diodes to make displays and other light sources is a possibility, e.g. "QD-LED" displays, and "QD-WLED" (White LED). Quantum dots are useful for displays, due to light-emitting in very specific Gaussian distributions.[10] A display that accurately renders the colours that the human eye can perceive is the result. Quantum dots require very little power, since they are not colour-filtered. With the discovery of "white-light emitting" QD, general solid-state lighting applications are achievable. A liquid crystal display (LCD) is powered by a single fluorescent lamp, that is colour-filtered to produce red, green, and blue pixels. Displays that intrinsically produce monochromatic light are more efficient, since they are visibly better.

QUANTUM WIRE

A quantum wire is an electrically conducting wire, in which quantum effects are affecting transport properties as per the condensed matter physics. The confinement of conduction electrons is in the transverse direction of the wire. Because of this transverse confinement, the transverse energy is quantized into a series of separate values. Exact calculation of the transverse energies of the confined electrons has to be performed to calculate a wire's resistance unlike formula for calculating the electrical resistivity of a wire. The quantization of electron energy and also the resistance is quantized for a quantum wire.

The quantization is inversely proportional to the diameter of the nanowire of a given material. It depends on the electronic properties, the effective mass of the electrons. It depends on the conduction electron interaction with the atoms within a given material. Semiconductors show conductance quantization for large wires having transverse dimensions of 100 nm because the electronic modes due to confinement are spatially extended. Their Fermi wavelengths[11] are large and have low energy separations. They can be resolved at cryogenic temperature (few kelvins) only, where the thermal excitation energy is lower than the inter-mode energy separation.

For metals, quantization of the lowest energy states is observed only for atomic wires. Since their wavelength is small, at room temperature, a very large energy separation occurs. This makes resistance quantization observation perfect.

Carbon Nanotubes as Quantum Wires

Quantum wires in limited quantities are made from metallic carbon nanotubes. They have high electrical conductivity (due to a high mobility), light weight, small diameter, low chemical reactivity, and high tensile strength except the negative economic angle. It is possible to create macroscopic quantum wires with a rope of carbon nanotubes. Quantum tunnelling will allow electrons to jump from strand to strand and there is no need for any single fibre to travel the entire length. Quantum wires for commercial uses are the next step in nanotechnology.

QUANTUM WELL

A quantum well is a potential well that confines particles and forces them to occupy a planar region to move them in two dimensions instead of the usual three-dimensional movements. The effects of quantum confinement take place when the quantum well thickness becomes comparable to the de Broglie wavelength[12] of the carriers (generally electrons and holes). This leads to energy levels called "energy sub bands" or discrete energy values.

FABRICATION

Quantum wells are formed in semiconductors using a material like gallium arsenide sandwiched between two layers of a material with a wider band gap, like aluminum arsenide. These structures are grown by molecular beam epitaxy or chemical vapour deposition with control of the layer thickness down to monolayers.

APPLICATIONS

Electrons in quantum wells have a sharper density of states due to their quasi two-dimensional nature than bulk materials. Because of that, quantum wells are widely used in

1. diode lasers, specifically blue lasers.
2. production of HEMTs (high electron mobility transistors), used in low-noise electronics.
3. quantum well infrared photodetectors used for infrared imaging.

By doping either the well or the barrier of a quantum well with donor impurities, a two-dimensional electron gas (2DEG) can be formed. This quasi two-dimensional system has interesting properties at low temperature like the quantum Hall effect[13] seen at high magnetic fields. Acceptor dopants can produce a two-dimensional hole gas (2DHG).

QUANTUM POINT CONTACT (QPC)

A quantum point contact (QPC) is a narrow constriction between two wide electrically-conducting regions, of a width comparable to the electronic wavelength (nano- to micrometre). Quantum point contacts were noted in 1988.

FABRICATION

For fabricating a QPC, pulling apart, in a break junction a piece of conductor until it breaks, is the methodology adopted. The breaking point forms the point contact. Quantum point contacts are formed in 2-dimensional electron gases (2DEG), e.g. in GaAs/AlGaAs[14] heterostructures, in a more controlled way. For creating different types of conducting regions like quantum dots and quantum point contacts in the plane of the 2DEG, the following methods are followed. It is done by applying a voltage to suitably shaped gate electrode and also by depleting electron gas locally.

Alternatively creating a point contact is done by positioning an STM-tip close to the surface of a conductor.

PROPERTIES

A quantum point contact is a constriction in the transverse direction which presents a resistance to the motion of electrons in geometrical terms. (Applying a voltage *V* across the point contact, a current I will flow, the size given by $I = GV$, where G is the conductance of the contact. This formula resembles Ohm's law for macroscopic resistors, but requires a quantum mechanical point of view).

APPLICATIONS

For studying fundamentals of charge transport in mesoscopic conductors,[15] quantum point contacts can also be used as extremely sensible charge detectors. Since the conductance through the contact strongly depends on the size of the constriction, any potential fluctuation (for instance, created by other electrons) nearby can influence the current through the QPC. Detection of single electrons can be made. In view of quantum computation in solid-state systems, QPCs can be used as read-out devices for the state of a qubit.

NANOCRYSTALS

Nanocrystal is defined as "nanometre-scale crystallite" or a crystalline form of a solid material in nanoscale dimension. This means that dimensionally the material is of a nanoscale structure and the arrangement of the atoms within the structure is in crystalline type of order. The presence of nanocrystals of various solids has been known for more than half a century and their images were noted through electron microscope in the 1950s itself. But, their very active surfaces, like biomolecules or polymers of the same dimension, which are difficult to handle, is an important reason for their unpopularity until recently.

The nanocrystal is of fundamental interest to scientists. It is also a bridge between the world of molecules and that of crystalline solids. So, nanocrystal lies clearly within the interdisciplinary world of "nanoscience." The role of nanocrystals within a new nanotechnology is projected and needs further work up before being applied in practice. Obviously, various metallic, magnetic and semiconductor nanocrystals are connected with research programmes for a futuristic nanoelectronics like computer circuits, sensors, bio-nanoelements, and applications of similar nature.

Gold is the archetypal[16] metallic element meaning, it is the original metal after which other similar metals are studied. It is the prototype and the most commonly investigated of colloidal metals since 1856 in modern science. But it was used in most of the older civilizations including Indian civilization and needs more research regarding its utility in those situations. It has a role in critical microelectronics and relays for a longer period of time.

Gold is a noble metal which means:

1. it rarely forms stable compounds with most of the substances that it is exposed to,

2. it can form a range of alloys with other metals and

3. it can form a protective plated coating on an even wider range of important (nano-) materials.

It has also been described to have a "simple nature" denoting bonding and electronic structure, and could be used as a prototype to illuminate many aspects of nanometals generally. So, it is usual to start with gold and silver and use them as reference materials for alloys and bimetallic (coated or plated) metals generally in metallic science.

Gold nanocrystals are being used as an important component of commercial catalysts developed in Japan for air-purification. Commercial applications of the thiolate-protected metal nanocrystals are developing more slowly, in spite of its application in nanoelectronics or sensing devices are projected widely with a big expectation.

The synthesis of CdTe nanocrystals (cadmium telluride) is much easier. Complex structures like nanowires and nanosheets are created out of spherical CdTe nanocrystals. The production of these crystals are scaled up and more work is on in that line.

For biological applications, brightly luminescent, water-soluble nanocrystals with a feasible surface chemistry are required. Biologists are using semiconductor nanocrystals for advanced cell imaging and trying to study the cells effectively.

Reliable methods of synthesis of any kind of metals and semiconductors are available. Those methods provide nanocrystals with the desired size, shape, and composition. Surface functionalization of nanocrystals on demand is the area of intense research. Application-related considerations drive the direction of progress in the nanocrystals field.

Semiconductor nanocrystals will be widely used as components of solar cells and also to be widely accepted imaging agents in biology within a period of ten years. The hybrid structures utilizing different functional components, like magnetic and luminescent nanocrystals are in the developmental stages.

Nanocrystals with complex shapes and composition are currently under investigation in various applications as potential materials. CdSe nanorods and CdTe tetrapods are shown to increase the efficiency of polymer-based solar cells. They are added as additives in the polymer layer. Larger efficiencies are expected when more and more complex nanostructures are incorporated. Polarized displays based on aligned nanorod arrays are of significance related to future application. Advantages of semiconductor nanorods as fluorescence labels in biological applications are also noted. In the fields of photonics and nanoelectronics, newer applications of these materials are expected.

LIMITATIONS

1. The toxicity of the materials which form nanocrystals are important limitation of these materials usage. It can be made to be reduced or eliminated by the development of non-toxic, or at least less toxic types of nanocrystals.

2. The fabrication of nanocrystals yields samples with a certain distribution of sizes and shapes and not as required by the manufacturer's specifications. The development of computational tools that model the growth process is the need of the hour.

Using the nanocrystals as a basic material and bottom-up self assembly, development of complex controllable geometries over large areas is of tremendous importance in nanocrystal field of science. This is an essential part of implementing shape-controlled nanocrystals in diverse applications. By using controlled assembly, the products developed have been useful in diverse applications. A self-assembly approach is useful to create complex systems with millions or even billions of nano-components. This will also help in the understanding and controlling the self-assembly of shape-controlled nanocrystals.

NANOCRYSTAL SOLAR CELL

Nanocrystal solar cells or quantum dot solar cells are solar cells based on a silicon substrate with a coating of nanocrystals. Instead of expensive molecular beam epitaxy processes, quantum dot creation can be done by colloidal synthesis, a more cost-effective manufacturing process. A thin film of nanocrystals is formed by a process known as "spin-coating". By placing an amount of the quantum dot solution onto a flat substrate, and then rotating very quickly, it is produced. The solution spreads out uniformly and the substrate is spun till the expected thickness is achieved.

Quantum dot-based photovoltaic cells based around the dye-sensitized colloidal films were found to be converting incident light energy to electrical energy. This is using the low-cost materials and so, this is a form of commercially viable/affordable renewable energy sources.

In future, quantum dot-based photovoltaics can offer advantages, such as mechanical flexibility (quantum dot-polymer composite photovoltaic) as well as low cost, clean power generation.

Lead selenide (PbSe) semiconductor, cadmiu telluride (CdTe), which has been used in the production of "classic" solar cells and other materials are under research.

ENDNOTES

1. In the physics field of semiconductors and insulators, the conduction band is the range of electron energy, higher than that of the valence band, sufficient to make the electrons free to accelerate under the influence of an applied electric field and thus constitute an electric current. Semiconductors may cross this conduction band when they are excited.

2. In solids, the valence band is the highest range of electron energies where electrons are normally present at zero. In semiconductors and insulators, there is a band gap above the valence band, followed by a conduction band. The hole is an empty state that allows electrons in the valence band some degree of freedom.

3. An exciton is a bound state of an electron and an imaginary particle called an electron hole in an insulator or semiconductor, and such is a Coulomb-correlated electron-hole pair. It is an elementary excitation, or a quasiparticle of a solid. In current research, the bound electron and hole pairs (excitons) provide a means to transport energy without transporting net charge.

4. A quantum bit or qubit is a unit of quantum information. That information is described by a state vector in a two-level quantum-mechanical system, which is formally equivalent to a two-dimensional vector space over the complex numbers.

5. Exciton Bohr Radius   This is the natural physical separation in a crystal between an electron in the conduction band and the hole it leaves behind in the valence band. The size of this radius controls how large a crystal must be before its energy bands can be treated as continuous. Therefore, the Exciton Bohr Radius can rightly be said to define whether a crystal can be called a semiconductor quantum dot, or simply a bulk semiconductor.

6. In atomic physics, Hund's rules refer to a simple set of rules which is used to determine the term symbol that corresponds to the ground state of a multi-electron atom.

7. Blue shift is the shortening of a transmitted signal's wavelength, and/or an increase in its frequency, due to the Doppler effect, which means the object is moving toward the observer. The shorter-wavelength end of the optical spectrum is the blue (or violet) end and hence the name. When visible light is compacted in wavelength, it is shifted towards the "blue" end of the spectrum, and said to be "blue-shifted". As the longer-wavelength end of the visible electromagnetic spectrum is red, the opposite effect, of a lengthening of a signal's wavelength, is

referred to as red shifting.

8. Stranski–Krastanov growth (SK growth, also Stransky–Krastanov or Stranski–Krastanow) is one of the three primary modes by which thin films grow epitaxially at a crystal surface or interface.

9. In physics, the increased resistance at small bias voltages of an electronic device comprising at least one low-capacitance tunnel junction is called as a Coulomb blockade effect named after Charles-Augustin de Coulomb.

10. The normal distribution is called the Gaussian distribution, is an important family of continuous probability distributions, applicable in many fields.

11. The Fermi wavelength is wavelength of carriers that dominate electrical transport.

12. The de Broglie relations show that the wavelength is inversely proportional to the momentum of a particle and that the frequency is directly proportional to the particle's kinetic energy.

13. The quantum Hall effect (or integer quantum Hall effect) is a quantum-mechanical version of the Hall effect, observed in two-dimensional electron systems subjected to low temperatures and strong magnetic fields.

14. The $Al_xGa_{1-x}As/GaAs$ heterostructure system is potentially useful material for high-speed digital, high-frequency microwave, and electro-optic device applications.

15. Mesoscopic conductors constitute ideal systems for the investigation of quasiparicle entanglement.

16. Archetypal is an original model or type after which other similar things is patterned; a prototype.

REVIEW QUESTIONS

1. Describe in detail about quantum dots.

2. Describe in detail about application of quantum dots.

3. Describe in detail about nanocrystals.

4. Write short notes on:

    i.   Nanophysics

    ii.   Quantum confinement in semiconductors

    iii. Optical properties of quantum dots

    iv   Fabrication of quantum dots

    v.   Quantum wire

vi.   Quantum well

vii. Quantum point contact

viii. Nanocrystal solar cell

REFERENCES

1. Bandyopadhyay, S. and Miller, A.E. (2001). "Electrochemically self-assembled ordered nanostructure arrays: Quantum dots, dashes, and wires." In: Nalwa, H.S. (ed.). *Handbook of Advanced Electronic and Photonic Materials and Devices*. Academic Press, San Diego.

2. Delerue, C. and Lannoo, M. (2004). *Nanostructures: Theory and Modelling*. Springer. p.47.

3. Fahlman, B.D. (2007). *Materials Chemistry*. Vol. 1. Springer, Mount Pleasant, MI. pp. 282–283.

4. Murray, C.B., Kagan, C.R. and Bawendi, M.G. (2000). "Synthesis and characterization of monodisperse nanocrystals and close-packed nanocrystal assemblies." *Annual Review of Materials Research*. 30 (1): 545–610.

5. O'Regan, B. and Gratzel, M. (1991). "A low-cost, high efficiency solar cell based on dye-sensitized colloidal TiO_2 films." *Nature*. (353): 737–740.

6. Reed, M.A. (1993). "Quantum dots." *Scientific American*. 268 (1): 118.

7. Silbey, Robert, J. Alberty, Robert, A. and Bawendi, Moungi, G. (2005). *Physical Chemistry*, 4th edn. John Wiley and Sons. p.835.

8. Thomas Engel. (2006). *Quantum Chemistry and Spectroscopy*. Pearson Education, pp. 75–76.

Chapter 4
Nanoelectronics

INTRODUCTION

Nanoelectronics can be defined as the use of nanotechnology on electronic components, especially transistors as well as in the research, designing and manufacturing of components useful in this field. The common definition of nanotechnology, "technology utilizing matter less than 100 nm size" is not strictly applicable to nanoelectronics. It usually refers to transistor devices that are so small. Their inter-atomic interactions and quantum mechanical properties need to be studied extensively. As a result, present transistors like Pentium 4 Processors from Intel do not fall under this category, in spite of these devices being manufactured under 90 nm or 65 nm technology.

Nanoelectronics is nicknamed as disruptive technology due to its differences from traditional transistors. Hybrid molecular/semiconductor electronics, one-dimensional nanotubes/nanowires, or advanced molecular electronics are some of the areas of nanoelectronics. The sub-voltage and deep sub-voltage nanoelectronics are specific and important fields of research and development. The appearance of new ICs operating at near theoretical limit in relation to fundamental, technological, design, methodological, architectural and algorithmic definitions on energy consumption per 1 bit processing is inevitable.

The above fields of nanoelectronics holds immense promises for the future and is still under development. So, it will take some time before it can be used for manufacturing.

APPROACHES TO NANOELECTRONICS

Nanofabrication

Single electron transistors, having transistor operation based on a single electron, are the basic materials fabricated under this field. Nanoelectromechanical systems are also included in nanoelectronics. Nanofabrication can be used to construct ultradense parallel arrays of nanowires, instead of synthesizing nanowires individually.

Nanomaterials Electronics

The uniform and symmetrical structure of nanotubes, being small, allow more transistors to be packed into a single chip. They allow a higher electron mobility (faster electron movement in the material), a higher dielectric constant (faster frequency), and a symmetrical electron/hole characteristic. Nanoparticles can also be used as quantum dots.

Molecular Electronics

Single-molecule devices are the expected products. Designing the device components to construct a larger structure or even a complete system on their own by using molecular self-assembly is the expected line of planning. This will be very useful for computing. This may even completely replace present FPGA (Field-programmable gate array) technology in future.

Molecular electronics is a new technology in its infant stage. It will be a truly atomic scale electronic system of the future. A molecular level diode/transistor synthesis by organic chemistry was proposed in "Molecules for Memory, Logic and Amplification."[1] A model system with a spiro carbon structure giving a molecular diode at nanometer scale which could be connected by polythiophene molecular wires was proposed. The possibility of developing such system which will work is projected for future.

Other Approaches

Study of the transport of ions in nanoscale systems is nanoionics. Study of the behaviour of light on the nanoscale is nanophotonics and this field has the goal of developing devices that take advantage of this behaviour.

Nanoelectronic Devices

Radios   Using carbon nanotubes, nanoradios have been developed.

Computers   Nanoelectronics aims at making computer processors more powerful than with conventional semiconductor fabrication techniques. New forms of nanolithography, the use of nanomaterials like nanowires or small molecules in place of traditional CMOS[2] components are researched. Field effect transistors are created using semiconducting carbon nanotubes as well as using heterostructured semiconductor nanowires.

Energy production   Using nanowires and other nanostructured materials to create cheaper and more efficient solar cells than with conventional planar silicon solar cells is an ongoing research. Invention of more efficient solar energy would fulfil the global energy needs.

Bio-nanogenerators (nanoscale electrochemical devices) are *in vivo* energy production devices using blood glucose in a living body and are also in exploratory stage.

Medical diagnostics   Constructing nanoelectronic devices to detect the concentrations of biomolecules in real time for use as medical diagnostics in nanomedicine is of great interest and is an ongoing activity. Interaction of nanoelectronic devices with single cells for use in basic biological research is also being developed and these are called nanosensors. Miniaturization in nanoelectronics towards *in vivo* proteomic sensing allows new approaches for health monitoring, surveillance, and defence technology.

NANOCIRCUITRY

INTRODUCTION

Nanocircuits are defined as electrical circuits working on the scale of nanometres. One nanometre can be equalized to a row of 10 hydrogen atoms. As the circuits are becoming smaller, more circuits can be made to fit on a computer chip. More complex functions using less power and also at a faster speed can be envisaged.

APPROACHES OF NANOCIRCUITRY

Single-electron transistors, quantum dot cellular automata and nanoscale crossbar latches are the areas in which nanocircuitry has been tried to be utilized. Incorporation of nanomaterials to improve the design of MOSFETs[3] will be the basis for practical applications. MOSFET (Figure 4.1) is currently forming the basis of most analog and digital circuit designs, the scaling of which drives Moore's law. The usage of pre-fabricated multiple nanowires for the channel increases reliability and reduces production costs. In this way, large-volume printing processes can be used to deposit the nanowires at a lower temperature than conventional fabrication procedures. Due to the lower temperature deposition, a wider variety of materials such as polymers can be used as the carrier substrate for the transistors than conventional ones. This opens nanocircuitry to flexible electronics applications such as electronic paper, bendable flat panel displays and ide-area solar cells.

Figure 4.1   MOSFETs

MOORE'S LAW AND NANOCIRCUITRY

For understanding nanocircuits, one of the most fundamental concepts available is the formulation of Moore's Law. This concept was thought of when Intel co-founder Gordon Moore was working about the cost of transistors and trying to fit more onto one chip. The law relates that the number of transistors that can be fabricated on a silicon-integrated circuit and therefore the computing speed of such a circuit—are doubling every 18 to 24 months. As a corollary, the more transistors when fit on a circuit, the faster will be the computer. Scientists and engineers are trying to produce these nanocircuits for this purpose. They are trying to fit in millions and perhaps even billions of transistors onto a chip. This may sound good, but plenty of problems will arise when so many transistors are packed together. Being so tiny, have more problems (especially many defects) than larger circuits. Nanoscale circuits are more sensitive to temperature changes, cosmic rays and electromagnetic interference than presently available circuits. Packing more transistors onto a chip will have the following effects.

1. produce phenomena such as stray signals on the chip

2. create the need to dissipate the heat from so many closely packed devices

3. difficulty in creating the devices in the first place

These will halt or severely slow down the progress. A time will come when the cost of making circuits even smaller will be more and the speed of computers will reach a maximum. Because of this achievement Moore's law will not hold forever and will soon reach a peak. In other words, the utility of Moore's law will be limited after some time and it is time-bound and will lose its applicability in the long run.

METHODS OF PRODUCTION

In producing nanocircuits, various steps are involved.

1. The first part of nanocircuit organization begins with transistors. Now, only silicon-based transistors are used. Transistors are an integral part of circuits. Their role is controlling the flow of electricity and transforming weak electrical signals to strong ones. They also control electric current by turning it on or off, or even amplifying signals. Silicon is used as a transistor in circuits because it can easily be switched between conducting and nonconducting states. In nanoelectronics, transistors might be formed out of organic molecules or nanoscale inorganic structures. Semiconductors, which are part of transistors, are also expected to be made of organic molecules in the nanostate.

2. The second aspect of nanocircuit organization is interconnection. This involves logical and mathematical operations. The wires linking the

transistors together make this possible. In nanocircuits, nanotubes and other wires as narrow as one nanometre are utilized to link transistors together. Nanowires are made from carbon nanotubes. Until a few years ago, transistors and nanowires were put together to produce the circuit. A nanowire with transistors embedded in it is produced by scientists. A nanowire, 10,000 times thinner than a sheet of paper containing a string of transistors, was produced by Charles Lieber and his team of Harvard University in 2004. Since transistors and nanowires are already pre-wired while being fabricated, the difficult task of trying to connect transistors together with nanowires at nanoscale, can be avoided.

3. The last part of nanocircuit organization is architecture. The way in which the transistors are interconnected is architecture. The architecture helps to decide how the circuit can plug into a computer or other system and operate independently of the lower-level details. With nanocircuits being so small, nanocircuits are prone for error and defects. A way to get around these lacunae has been devised by scientists. Their architecture combines circuits that have redundant logic gates and interconnections. They have the ability to reconfigure structures at several levels on a chip. The redundancy makes the circuit to identify problems on its own and reconfigure itself so that the circuit can avoid more problems. It also allows for errors within the logic gate and in spite of that, it works properly without giving a wrong result.

This can be christened as self-correction mechanism of nanocircuit or self-regulating mechanism of nanocircuit. This is a unique property of nanocircuits and needs more study for application in other areas of materials science and bioscience. Whether, the same mechanism holds good related to improvement in neurological defects after a neurological illness which affects the functioning of the patient, needs an in-depth study for better understanding of the neurological processes in disease and illness and also for therapeutic applications. The author is of the view that this will create a new science in neurology called neurocircuitry like nanocircuitry. Neuro-circuit and neural circuit are already studied especially in animals and flies and even drawn for specific functions. This theory is based on self correcting mechanisms of neuron and its applications in illness and improvement.

Indian scientists have developed the world's smallest transistor which will be used for nanocircuits recently. The transistor is created entirely from carbon nanotubes. These transistors have two different branches and they meet at a single point, giving it a Y shape. Current can flow throughout both branches and is controlled by a third branch that is responsible for the voltage on or off. This new breakthrough makes nanocircuits's name true as they are made entirely from nanotubes. Prior to this discovery, logic circuits used nanotubes, but to control the flow of current, metal gates were used.

POTENTIAL APPLICATIONS

The biggest potential applications of nanocircuits are with computers and electronics. Research is on to make computers faster. Hybrids of micro and nano, i.e., silicon with a nanocore will produce a high-density computer memory that retains its contents forever. Nanocircuit design will begin with the chip, unlike conventional circuit design, which proceeds from blueprint to photographic pattern to chip. Nanocircuit is a haphazard jumble of around 1024 components and wires. In this complicated circuit, all the components will not work like conventional circuit. But there will be intercircuit adjustment which will ultimately make it a useful device. The bottom-up approach will have to be adopted, instead of the traditional top-down approach, due to the smaller size of these nanocircuits. Nanocircuits will be more defective at nanolevel and faulty because of their compactness and not everything in the circuit is expected to work. All the essential components of nanocircuits like transistors, logic gates and diodes have been created. They were all constructed from organic molecules, carbon nanotubes and nanowire semiconductors. Once the ways to eliminate the errors are made that come with such a small device and nanocircuits, it will become the way of all electronics. At one stage, a limit as to how small nanocircuits can become, will be reached. Ultimately, computers and electronics will end up in their equilibrium speeds.

ECONOMIC IMPACT

When the size of the circuits become smaller, the expenses in producing these will be on the rise. A fabrication facility for making nanocircuit could cost very dearly to the manufacturers. More time and effort spent and the difficulty of producing such circuits will automatically increase the cost of the final products. The fabrication plant is expected to initially create a raw nanocircuit which is a lump of material with a handful of wires sticking out; its functioning will be limited. Eventually, the theory of Moore's Law will reach an equilibrium with the fabrication methods currently used. In the long run, circuits in future manufacturing methods are expected to be faster and smaller without creating any severe problems. So, for producing better nanocircuits, the manufacturing cost will increase further because of expenses to develop new fabrication methods and ways of designing faster, better nanocircuits. Till then, the end users would have to accept the present products be the best in the market. Although nanocircuits may have problems, the research on advanced versions and production of the same will be an ongoing process among the manufacturers to cope up with the competition and so the consumers will have to bear the trial expenses also to utilize such novel products initially.

MOLECULAR LOGIC GATE

A molecular logic gate is a molecule that performs a logical operation on one or more logic inputs and produces a single logic output. Several prototypes now exist due to consistent research. Their potential utility in simple arithmetic, make these molecular machines as moleculators.

Molecular logic gates work with input signals based on chemical processes and with output signals based on spectroscopy.

MOLECULAR ELECTRONICS

Molecular electronics or moletronics is an interdisciplinary concept which encompasses basic sciences namely physics, chemistry and materials science. Molecular building blocks used for the fabrication of electronic components, both passive (e.g. resistive wires) and active (e.g. transistors) is the unifying feature of molecular electronics. Also, biological or organic molecules are used for fabricating the electronic materials with novel electronic, optical, or magnetic properties, applications which include polymer light-emitting diodes, conductive polymer sensors, pyroelectric plastics, and potentially, molecular computational devices. The concept of molecular electronics has the prospect of size reduction in electronics offered by molecular-level control of properties and so it is interesting to the scientific community. Molecular electronics is a way to extend Moore's law apart from small-scale conventional silicon integrated circuits.

Molecular electronics is classified into two related but separate subdisciplines.

1. Molecular materials are used in electronics due to the properties of the molecules which can affect the bulk properties of a material.
2. Molecular scale electronics, focuses on single-molecule applications.

In 1940, Robert Mulliken and Albert Szent-Gyorgi developed the study of charge transfer and energy transfer in molecules using "donor–acceptor" systems concept. Mark Ratner and Ari Aviram illustrated a theoretical molecular rectifier in 1974. In 1988, Aviram described in detail a theoretical single-molecule field-effect transistor. Further concepts, including single-molecule logic gates were proposed by Forrest Carter of the Naval Research Laboratory.

The direct measurement of the electronic characteristics of individual molecules needs the development of methods for making molecular-scale electrical contacts. Mark Reed and his co-workers have reported the meaning of the conductance of a single molecule in 1977. The theoretical predictions of the early workers have been mostly confirmed. The field is now in a progressive stage of development. Developing circuits that are "grown" may seem far-fetched, but research shows that a "biocomputer" might be possible in 25 years in biological based electronics. From living plants, researchers have isolated "photosynthetic reaction centres" (PRCs).

These microminiature photovoltaic systems change their state when exposed to light, and the states can be measured.

Molecular electronic processes also manifest on a macroscale but mostly in nanoscale. They include quantum tunnelling, negative resistance, phonon-assisted hopping, and polarons. Macro-scale active devices were described decades before molecular-scale ones.

> In 1974, John McGinness and his co-workers described the "first experimental demonstration of an operating molecular electronic device" which was a voltage-controlled switch. This device used DOPA melanin, an oxidized mixed polymer of polyacetylene, polypyrrole, and polyaniline as its active element. The "ON" state of this switch exhibited almost metallic conductivity.

Building Blocks

The possible building blocks of molecular electronics are as follows:

Charge transfer complexes   The charge transfer complexes are organic compounds with high-conductivity. In 1954, low resistivity (8 ohm/cm) charge transfer complexes were reported. Metallic conductivity (1970) and superconductivity (1980) of tetrathiafulvalene salts were demonstrated.

Conductive polymers   The linear-backbone "polymer blacks" like polyacetylene, polypyrrole, and polyaniline and their copolymers are the main class of conductive polymers. They are also known as melanin. Iodine-doped oxidized polypyrrole blacks with resistivity as low as 1 ohm/cm were reported in 1963 by the Australian De Weiss and his co-workers and resistances as low as 0.03 ohm/cm were reported later. Organic molecules were previously considered as insulators or weakly conducting semiconductors except charge transfer complexes, which are acting as superconductors.

In 1977, equivalent high conductivity as in charge transfer complexes was reported in oxidized and iodine-doped polyacetylene by Shirakawa, Heeger, and MacDiarmid. They got the Nobel Prize in chemistry in 2000 for "The discovery and development of conductive polymers".

From graphite to C_{60}   In polymers, classical organic molecules are composed of both carbon and hydrogen and at times, compounds such as nitrogen, chlorine or sulphur. They are derived from petrol and can be synthesized in bulk. Most of these molecules are insulating related to electricity if their length exceeds a few nanometres. Naturally occurring carbon is a conductor. Graphite, recovered from coal or encountered naturally is also a conductor. Graphite is a semi-metal, a category in between metals and semiconductors.

It has a layered structure, each sheet being one atom thick. In between each sheet, the interactions are weak and so, manual cleavage is easier.

Making the graphite sheet as nanometre-sized objects is a dream at the moment. Extremely small graphitic objects as single molecules were tried to be fabricated by twentieth century. Coined buckminsterfullerenes, buckyballs or C_{60}, the clusters retained some properties of graphite, such as conductivity.

Theory

The theory of single molecule devices deals with the system which is an open quantum system in nonequilibrium and driven by voltage. They are studied as follows:

1. In the low-bias voltage regime, the nonequilibrium nature of the molecular junction is ignored and the current–voltage characteristics of the device are calculated using the equilibrium electronic structure of the system.

2. In stronger bias regime, a more sophisticated treatment is required, as there is no longer a variational principle.

    i. In the elastic tunnelling case (where the passing electron does not exchange energy with the system), the formalism of Rolf Landauer[4] can be used to calculate the transmission through the system as a function of bias voltage, and hence the current.

    ii. In inelastic tunnelling, Meir-Wingreen formulation[5] has been used to examine the more difficult and interesting cases where the transient electron exchanges energy with the molecular system (for example, through electron–phonon coupling or electronic excitations).

NANOWIRE

A nanowire is a wire with a diameter of the order of a nanometre (10^{-9} m). Also, nanowires can be defined as structures that have a lateral size constrained to tens of nanometres or less and an unconstrained longitudinal size. At these scales, quantum mechanical effects are important—hence such wires are also known as "quantum wires". Many different types of nanowires exist, including metallic (e.g. Ni, Pt, Au), semiconducting (e.g. Si, InP, GaN, etc.), and insulating (e.g. SiO_2, TiO_2). The nanowires could be used, in the near future, to link the tiny components into extremely small circuits. Using nanotechnology, such components could be created out of chemical compounds.

NANOPHOTONICS

Nanophotonics is the study of the behaviour of light on the nanometre scale. The ability to fabricate the devices in nanoscale that has been developed recently provided the catalyst for this area of study. Nanophotonics scientists are experimenting with different ways to generate and manipulate light using ultrasmall, engineered structures at the nanoscale. Tiny gold particles called "nanostars" are being studied to ascertain how they interact with light. The study has the potential to revolutionize the telecommunications industry by providing low power, high speed, interference-free devices such as electro-optic and all-optical switches on a chip.

NANOIONICS

The study and application of phenomena, properties, effects and mechanisms of processes connected with fast ion transport (FIT) in all-solid-state nanoscale systems is called nanoionics. Fundamental properties of oxide ceramics at nanometre length scales and fast ion conductor (advanced superionic conductor) or electronic conductor heterostructures are the interesting topics in nanoionics. Potential applications of nanoionics in electrochemical devices (electrical double-layer devices) are in conversion and storage of energy, charge and information. In January 1992, A.L. Despotuli and V.I. Nikolaichik (Institute of Microelectronics Technology and High Purity Materials, Russian Academy of Sciences, Chernogolovka) first introduced the term and conception of nanoionics as a new branch of science.

There are two classes of solid-state ionic nanosystems which are fundamentally different nanoionics. They are:

1. nanosystems based on solids with low ionic conductivity

2. nanosystems based on advanced superionic conductors (alpha–AgI, rubidium silver iodide–family, etc.).

An important case of fast ionic conduction in solid states is in surface space-charge layer of ionic crystals. A significant role of boundary conditions with respect to ionic conductivity was first discovered and found as an anomalously high conduction in the Lil-Al_2O_3 two-phase system. Because a space-charge layer with specific properties has nanometre thickness, the effect is directly related to nanoionics (nanoionics-I). This is called Lehovec effect and has become the basis for the creation of a multitude of nanostructured fast ion conductors which are used in modern portable lithium batteries and fuel cells.

The examples of nanoionic devices are all-solid-state supercapacitors with fast ion transport at the functional heterojunctions (nanoionic supercapacitors), lithium batteries and fuel cells with nanostructured electrodes, nano-switches with quantized conductivity on the basis of fast ion conductors. These are well compatible with sub-voltage and deep-sub-voltage nanoelectronics and could find wide applications, e.g. as in autonomous micropower sources, Radio frequency identification (RFID), MEMS, smart dust, nanomorphic cell, other micro- and nanosystems, or reconfigurable memory cell arrays (computer data storage).

Being a branch of science and technology, nanoionics is unambiguously defined by its own objects (nanostructures with FIT), subject matters (properties, phenomena, effects, mechanisms of processes, and applications connected with FIT at nanoscale), methods (interface design in nanosystems of superionic conductors), and criterion (R/L ~1, where R is the nanosize(s) of device structure, and L is the characteristic length) on which the properties, characteristics, and other parameters (connected with FIT) change drastically.

Nanoionics-I and nanoionics-II differ from each other in the design of interfaces. The role of boundaries in nanoionics-I is the creation of conditions for high concentrations of charged defects (vacancies and interstices) in a disordered space-charge layer. But in nanoionics-II, it is necessary to conserve the original highly ionic conductive crystal structures of advanced superionic conductors at ordered (lattice-matched) heteroboundaries.

The International Technology Roadmap for Semiconductors (ITRS), relate nanoionics-based resistive switching memories to the category of "emerging research devices" ("ionic memory"). The area of close intersection of nanoelectronics and nanoionics can be called nanoelionics. Future short-sized devices may be nanoionic, i.e., based on the fast ion transport at the nanoscale.

RECENT PROGRESS

Recent progress in nanotechnology and nanoscience has enabled both experimental and theoretical study of molecular electronics. The development of the scanning tunnelling microscope (STM) and the atomic force microscope (AFM) have made it possible for the manipulation of single-molecule electronics.

1. Using a mechanical break junction approach to connect two gold electrodes to a sulphur-terminated molecular wire was the first work.

2. Molecular electronics based on rotaxanes and catenanes was also developed.

3. Use of single-wall carbon nanotubes as field effect transistors was developed.

4. The Aviram–Ratner model is used as a molecular rectifier. Many rectifying molecules have been identified, and are expanding rapidly.

5. Supramolecular electronics is a new field that tackles electronics at a supramolecular level.

The determination of the resistance of a single molecule (both theoretical and experimental) is an important issue in molecular electronics. STM is used to determine the conductivity of a molecule. It is difficult to prform direct characterization by STM, since imaging at the molecular scale is often difficult.

ENDNOTES

1. "Molecules for memory, logic, and amplification." Ari. Aviram, *J. Am. Chem. Soc.*, 1988, 110 (17), pp. 5687–5692, August 1988.

2. Complementary metal–oxide–semiconductor (CMOS) is a major class of integrated circuits. CMOS technology is used in microprocessors, microcontrollers, static RAM and other digital logic circuits. CMOS technology is also used for a wide variety of analog circuits such as image sensors, data converters, and highly integrated transceivers for many types of communication.

3. The metal–oxide–semiconductor field-effect transistor (MOSFET, MOS-FET, or MOS FET) is a device used to amplify or switch electronic signals. It is by far the most common field effect transistor in both digital and analog circuits.

4. The Landauer Formula: Landauer Formula is useful in order to compute current in nanoscale devices. It is useful to compute transport in carbon nanotubes and silico nanowire transistors.

5. Meir-Wingreen formulation: A formula used for mesoscopic structures.

REVIEW QUESTIONS

1. Discuss about nanoelectronics.

2. What is nanocircuitry?

3. Write an account of molecular electronics.

4. Write short notes on:
   i. Nanoelectronics
   ii. Nanoelectronic devices
   iii. Approaches to nanoelectronics
   iv. Moore's law and nanocircuitry
   v. Potential applications of nanocircuitry
   vi. Methods of production of nanocircuitry
   vii. Mlecular logic gate
   viii. Conductive polymers
   ix. Nanowire

    x.   Nanophotonics

    xi.   Nanoionics

REFERENCES

1. Aviram, A. (1988). "Molecules for memory, logic, and amplification." *Journal of the American Chemical Society.* 110 (17): 5687–5692.

2. Aviram, A. and Ratner, M.A. (1974). "Molecular rectifier." *Chemical Physics Letters.* 29: 277.

3. Jensen, K., Weldon, J., Garcia, H. and Zettl, A. (2007). "Nanotube Radio." *Nano lett.* 7 (11): 3508–3511.

4. Petty, M.C., Bryce, M.R. and Bloor, D. (1995). *Introduction to Molecular Electronics.* Oxford University Press, New York. pp.1–25.

5. Saito, S. (1997). "Carbon nanotubes for next-generation electronics devices." *Science.* 278:77–78.

6. Tour, M. James *et al.* (1998). "Recent advances in molecular scale electronics." *Annals of the New York Academy of Sciences.* 852: 197–204.

Chapter 5
Molecular Machine

INTRODUCTION

A molecular machine can be defined as a unit that has a number of discrete molecular components, performing mechanical acts as output while responding to an input which is a specific stimulus. By this definition, it can be derived that the following criteria are needed to call a unit as a molecular machine:

1. a discrete number of molecular components,
2. performing mechanical-like movements (output)
3. in response to specific stimuli (input).

Molecular machine is applicable generally to molecules which have functions like that of ones at macroscopic level. This is a common term in nanotechnology. A number of highly complex molecular machines have been envisaged for constructing a molecular assembler ultimately.

Molecular systems that can shift a chemical or mechanical process away from equilibrium or basal state of reaction, represent a potentially important branch of chemistry and nanotechnology. By the above definition, these types of systems are a type of molecular machinery, as the gradient generated from this process is able to perform useful work.

HISTORY

There are two thought experiments, namely, Maxwell's Demon and Feynman's Ratchet (or Brownian ratchet) that form the historical basis for molecular machines.

The Brownian ratchet is a thought experiment about an apparent perpetual motion machine conceived by Richard Feynman in a physics lecture at the California Institute of Technology on May 11, 1962, as an illustration of the laws of thermodynamics. The simple machine, consisting of a paddlewheel and a ratchet, appears to be an example of a Maxwell's demon, able to extract useful work from random fluctuations in a system at thermal equilibrium.

A slightly different interpretation of Richard Feynman's ratchet is shown in Figure 5.1. It is a very small system of two paddles or gears connected by a rigid axle, these two paddles are kept at two different temperatures.

One of the gears (at T2) has a pawl for rectifying the system of motion and so the axle can move only in a clockwise rotation. While doing so it could lift a weight (m) upward upon ratcheting (a ratchet is more easily turned in one direction than another).

The paddle in box T1, if in a much hotter environment than the gear in box T2, the kinetic energy of the gas molecules hitting the paddle in T1 would be much higher than the gas molecules hitting the gear at T2. With lower kinetic energy of the gases in T2, very little resistance from the molecules on colliding with the gear in the opposite direction can be noted. The ratcheting allows for directionality. Slowly over time, the axle would rotate and the ratchet would lift the weght (m).

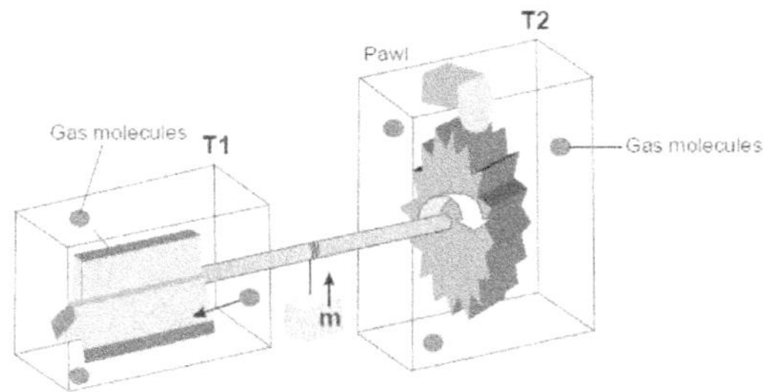

Figure 5.1 Schematic figure of Feynman's ratchet

On description, this system may appear to be a perpetual motion machine, but the key ingredient is the heat gradient within the system. This ratchet is within the frame of the second law of thermodynamics since this temperature gradient must be maintained by external source. Brownian motion of the gas particles provides the power to the machine. The temperature gradient is responsible for the machine to drive the system cyclically away from equilibrium. An interesting design concept in Feynman's ratchet is that random Brownian motion is not a negative force, but, harnessed and rectified as a positive force. Temperature gradients over molecular scale distances cannot be maintained due to molecular vibration which redistributes the energy to other parts of the molecule. Feynman's machine's useful work in lifting the mass is similar to using Brownian motion to power a molecular level machine. But how that power (or potential energy of the lifted weight, m) can be used to perform nanoscale tasks is yet to be analysed in practical terms.

MODERN INSIGHTS AND STUDIES

Harnessing molecular motion is a far more difficult process in molecular systems due to significant dynamic motions subject to the laws of Brownian mechanics (or Brownian motion), unlike macroscopic motion. Many machines at the macroscopic level operate in the gas phase and often air resistance is neglected, due to its insignificance. For a molecular system in a Brownian environment, molecular motion is similar "to walking in a hurricane, or swimming in molasses." The phenomenon of Brownian motion observed by Robert Brown, a botanist in 1827, was later explained by Albert Einstein in 1905. Einstein found that the Brownian motion, is a consequence of scale and not due to the nature of the surroundings. When thermal energy is applied to a molecule, it undergoes Brownian motion with the kinetic energy related to that temperature. When designing a molecular machine as like that of Feynman's strategy, it seems sensible to utilize Brownian motion to be harnessed and not to be fought.

Both macroscopic and molecular machines have movable parts. Macroscopic machines cannot be the same as molecular machines, and analogies between their design strategies will be misleading. The dynamics of macro and microscales are too different. To harness Brownian motion to make molecular level machines working is regulated by the Second law of thermodynamics.

Practically, harnessing Brownian motion is difficult but nature has provided blueprints for molecular motion performing useful work. Nature has created many useful structures. An example for such a structure is the cell membrane. Molecular systems are compartmentalized by cell membrane, thus creating nonequilibrium distributions. In such compartmentalization, lipophylic barriers use different mechanisms to power motion from one compartment to another.

TYPES OF MOLECULAR MACHINES

Based on the synthesis, molecular machines are of two types, namely molecular switches (or shuttles) and molecular motors. A switch influences a system as a function of state and a motor influences a system as a function of path. This is the major difference between the two. The function of a switch (or shuttle) creates an impression of translational motion. When a switch returns to its original position, mechanical effect is stopped and the energy is liberated to the system. For driving a system away from the equilibrium state repetitively or progressively, switches cannot use chemical energy. A motor can use chemical energy for this purpose.

Molecular machines fall under two broad categories namely synthetic and biological.

Synthetic Molecular Machines

A variety of simple molecular machines have been synthesized in chemistry. They may be a singe molecule. They are used for mechanically-interlocked molecular architectures, such as rotaxanes[1] and catenanes.[2] The following are some of the molecular machines described.

1. Molecular motors are molecules having capability of unidirectional rotation motion powered by external energy input like light or reaction. Molecular machines powered by light or reaction with other molecules have been synthesized.

2. A molecular propeller is a molecule that can propel fluids when rotated, due to its special shape like that of macroscopic propellers. Several molecular-scale blades attached at a certain pitch angle around the circumference of a nanoscale shaft is their design.

3. A molecular switch is a molecule that is reversibly shifted between two or more stable states. The molecules are shifted in response to changes like pH, light, temperature, an electrical current, micro-environment, or the presence of a ligand.

4. A molecular shuttle is a molecule that is capable of shuttling molecules or ions from one location to another. A rotaxane is a common molecular shuttle where the macrocycle can move between two sites or stations along the dumb-bell backbone.

5. Molecular tweezers are host molecules that are capable of holding the guest molecules between two arms. The open cavity of the molecular tweezers binds guest molecules using noncovalent bonding including hydrogen bonding, metal coordination, hydrophobic forces, van der Waals forces, $\pi-\pi$ interactions[3], and/or electrostatic effects. Molecular tweezers are constructed from DNA and are considered DNA machines.

6. A molecular sensor is a molecule that interacts with an analyte to produce a detectable change. Molecular sensors combine molecular recognition with some form of reporter to observe the presence of guest molecules.

7. A molecular logic gate is a molecule that performs a logical operation on one or more logic inputs and produces a single logic output. Compared to a molecular sensor, the molecular logic gate produces output for a particular combination of inputs.

Biological Molecular Machines

Molecular machines found within the biological systems are called biological molecular machines (Figure 5.2). The most complex molecular machines are within the cells.

These include motor proteins like myosin—responsible for muscle contraction, kinesin which moves cargo inside cells away from the nucleus along microtubules and dynein that produces the axonemal beating of cilia and flagella. These proteins are far more complex than any molecular machins so far artificially constructed.

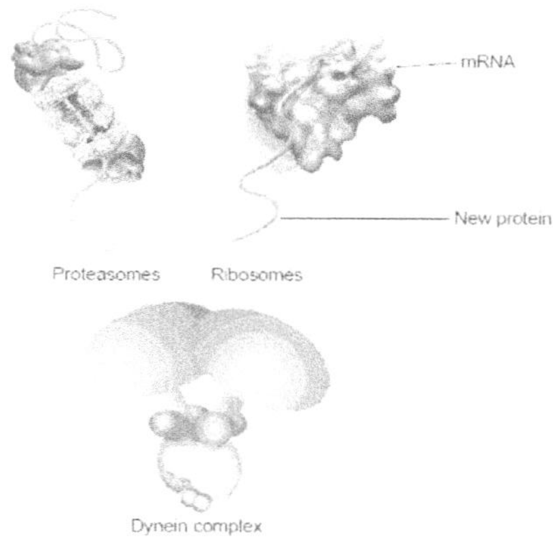

Figure 5.2 Some biological molecular machines

Theoretical Molecular Machines

Development of complex molecular machines is an active area of theoretical research. A number of molecules like molecular propellers are designed but no experimental studies are carried out due to the lack of methods to construct these molecules. These complex molecular machines form the basis of areas of nanotechnology, including molecular assembler.

MOLECULAR WIRES

Molecular wires or molecular nanowires are molecular-scale objects. They are the fundamental building blocks for molecular electronic devices and conduct electrical current. Their diameters are less than 3 nm and their bulk lengths may be extending to centimetres or more in a macroscopic scale.

Materials

Organic molecules form the basis of molecular wires. Highly conjugated systems produce higher conductivities. Alkane chains are necessary for understanding the basic charge transfer and tunnelling. DNA is a natural molecular wire. Prominent inorganic ones include polymeric materials such as $Li_2Mo_6Se_6$ and $Mo_6S_{9-x}I_x$, and single-molecule extended metal atom chains (EMACs) which comprise strings of late transition metal atoms directly bonded to each other. Molecular wires containing paramagnetic inorganic moieties can lead to observations of Kondo peaks which are interesting.

Structure

Unlike the usual nanowires (which are very thin crystals), the molecular nanowires are composed of repeating molecular units, either organic (e.g. DNA) or inorganic (e.g. $Mo_6S_{9-x}I_x$). The repeated units are the nucleotides with a backbone that is made of sugars and phosphate groups joined by ester bonds in the case of DNA. Attached to each sugar is one of four types of bases. In case of $Mo_6S_{9-x}I_x$ (Figure 5.3), the repeat units are $Mo_6S_{9-x}I_x$ clusters, that are joined together by sulphur or iodine bridges. Molecular nanowires often aggregate in solution into swatches or bundles. In the case of the Mo chalcogenide-halides, they grow in the form of ordered strands, in which the individual strands are linked by very weak van der Waals forces. But in EMAC, molecular wires consist of distinct molecules that do not aggregate and hence enable control of the exact ength of the wire at the atomic scale.

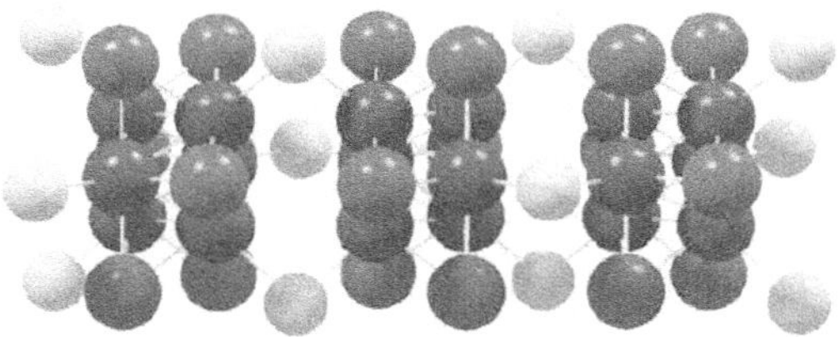

Figure 5.3 The structure of a Mo6S9-xIx molecular wire

Conduction of Electrons

Molecular wires conduct electricity. They have non-linear current–voltage characteristics and are not as simple as ohmic conductors. The conductance follows typical power law behaviour as a function of temperature or electric field, whichever is the greater, arising from their strong one-dimensional character. Numerous theoretical ideas have been used in an attempt to understand the conductivity of one-dimensional systems, where strong interactions between electrons lead to departures from normal metallic (Fermi liquid) behaviour. Effects caused by classical Coulomb repulsion, called Coulomb blockade, have also been found to be important in determining the properties of molecular wires (MWs).

Molecular Nanowires in Molecular Electronics

MWs, for connecting molecules together need to display some very important characteristics.

1. The connectors between elements need to be able to self-assemble by following well-defined routes and form reliable electrical contacts between them.

2. For self assembly of single molecules based complex circuit, to be in a reproducible way the recognitive ability of the connectors which join them is needed. They should be able to connect to diverse materials, such as gold metal surfaces (for connections to outside world), biomolecules (for nanosensors, nanoelectrodes, molecular switches) and most importantly, they must allow branching.

3. The connectors should be available in a pre-determined diameter and length. They should have covalent bonding to ensure reproducible transport and contact properties.

DNA-like molecules have specific molecular-scale recognition and can be used very effectively in molecular scaffold fabrication. Metal-coated DNA is electrically conducting but much too thick to connect to individual molecules. Thinner coated DNA lacks electronic connectivity and are not suitable for connecting molecular electronics elements. Some varieties of carbon nanotubes are conducting, and connectivity at their ends can be achieved by attachment of connecting groups. Unfortunately manufacturing CNTs with pre-determined properties is impossible at present, and the functionalized ends are typically non-conducting, limiting their usefulness as molecular connectors. Individual CNTs can be soldered in an electron microscope, but the contact is noncovalent and cannot be self-assembled. Recently, possible routes for the construction of larger functional circuits using $Mo_6S_{9-x}I_x$ MWs have been demonstrated either via gold nanoparticles as linkers or by direct connection to thiolated molecules. These two approaches may lead to different possible applications. The use of gold nanoparticles (GNPs) offers the possibility of branching and construction of larger circuits.

NANOELECTROMECHANICAL SYSTEMS

Nanoelectromechanical systems (NEMS) are like microelectromechanical systems (MEMS) but smaller in size. They have got the capacity to improve abilities to measure the small displacements and forces at a molecular scale. They are related to nanotechnology and nanomechanics. Two standard approaches are available and accepted to NEMS. They are

1. The top-down approach—"a set of tools designed to build a smaller set of tools." A millimetre sized factory that builds micrometre sized factories which in turn can build nanometre sized devices.

2. The bottom-up approach—putting together single atoms or molecules until a desired level of complexity and functionality has been achieved in a

device. Such an approach utilizes molecular self-assembly or mimic molecular biology systems.

The first very large scale integration (VLSI) NEMS device was demonstrated in 2000 by researchers from IBM.[4] Its premise was an array of AFM tips which can heat/sense a deformable substrate in order to function as a memory device.

The international technical roadmap for semiconductors (ITRS) contain NEMS memory as a new entry for the emerging research devices section in 2007. This is an indication that the semiconductor industry is actively considering the technology for implementation in the near future. A combination of these approaches can also be used, in which nanoscale molecules are integrated into a top-down framework, e.g. carbon nanotube nanomotor.

NANOMECHANICS

Nanomechanics is a branch of nanoscience studying the fundamental mechanical (elastic, thermal and kinetic) properties of physical systems at the nanometre scale. Nanomechanics has emerged on the cross-road of classical mechanics, solid-state physics, statistical mechanics, materials science, and quantum chemistry. As an area of nanoscience, nanomechanics provides a scientific foundation of nanotechnology. Often, nanomechanics is viewed as a branch of nanotechnology, i.e., an applied area with a focus on the mechanical properties of engineered nanostructures and nanosystems (systems with nanoscale components of importance). Examples of the latter include nanoparticles, nanopowders, nanowires, nanorods, nanoribbons, nanotubes, including carbon nanotubes and boron nitride nanotubes (BNNTs); nanoshells, nanomembranes, nanocoatings, nanocomposite/nanostructured materials, nanofluids (fluids with dispersed nanoparticles); nanomotors, etc.

Some of the well-established fields of nanomechanics are: nanomaterials, nanotribology[5] (friction, wear and contact mechanics at the nanoscale), nanoelectromechanical systems (NEMS) and nanofluidics.

As a fundamental science, nanomechanics is based on some empirical principles (basic observations):

1.   General mechanics principles
2.   Specific principles arising from the smallness of physical sizes of the subject of study or research.

General mechanics principles include:

◈   Energy and momentum conservation principles

◈   Variational Hamilton's principle

◈   Symmetry principles

Specific empirical principles of nanomechanics include:

◈   Discreteness of the subject (subject size becomes comparable with the interatomic lengths)

◈   Plurality, but finiteness of degrees of freedom in the subject

◈   Rise of thermal fluctuations

◈   Rise of configuration entropy

◈   Rise of quantum effects

These principles serve to provide a basic insight into novel mechanical properties of nanometre objects. Novelty is understood in the sense that these properties are not present in similar macroscale objects or much different from those properties (e.g. nanorods vs. usual macroscopic beam structures). In particular, smallness of the subject itself gives rise to various surface effects determined by higher surface-to-volume ratio of nanostructures and thus affects mechanoenergetic and thermal properties (melting point, heat capacitance, etc.) of nanostructures.

1. Discreteness serves a fundamental reason, for instance, for the dispersion of mechanical waves in solids and some special behaviour of basic elastomechanics solutions at small scales.

2. Plurality of degrees of freedom and the rise of thermal fluctuations are the reasons for thermal tunnelling of nanoparticles through potential barriers, as well as for the cross-diffusion of liquids and solids.

3. Smallness and thermal fluctuations provide the basic reasons of the Brownian motion of nanoparticles. Increased importance of thermal fluctuations and configuration entropy[6] at the nanoscale give rise to super elasticity, entropic elasticity (entropic forces) and other exotic types of elasticity of nanostructures.

4. Configuration entropy are of great interest in the context of self-organization and cooperative behaviour of open nanosystems.

5. Quantum effects determine forces of interaction between individual atoms in physical objects. They are introduced in nanomechanics by means of some averaged mathematical models called interatomic potentials.

Subsequent utilization of the interatomic potentials within the classical multibody dynamics provides deterministic mechanical models of nanostructures and systems at the atomic scale/resolution. Numerical methods of solution of these models are called molecular dynamics (MD), and sometimes molecular mechanics (especially, in relation to statically equilibrated (still) models). Non-deterministic numerical approaches include Monte-Carlo, Kinetic More-Carlo (KMC) and other methods. Contemporary numerical tools also include hybrid multiscale approaches allowing concurrent or sequential utilization of the atomistic scale methods (usually, MD) with the continuum (macro) scale methods (usually, FEM) within a single mathematical model. Development of these complex methods is a separate subject of applied mechanics research.

Quantum effects also determine novel electrical, optical and chemical properties of nanostructures and in nanoscience and nanotechnology, such as nanoelectronics, advanced energy systems and nanobiotechnology they find even greater attention and application.

Gears, bearings, and liquid lubricants can reduce friction in the macroscopic world, but the origins of friction for small devices such as micro- or nanoelectromechanical systems (NEMS) require other solutions. Despite the unprecedented accuracy by which these devices are nowadays designed and fabricated, their enormous surface–volume ratio leads to severe friction and wear issues, which dramatically reduce their applicability and lifetime. Traditional liquid lubricants become too viscous when confined in layers of molecular thickness. This situation has led to a number of proposals for ways to reduce friction on the nanoscale, such as superlubricity and thermolubricity.

NANOLITHOGRAPHY

Nanolithography otherwise also called as photolithography at the nanometre scale refers to the fabrication of nanometre-scale structures, meaning patterns with at least one lateral dimension between the size of an individual atom and approximately 100 nm. Nanolithography is used during the fabrication of leading-edge semiconductor integrated circuits or nanoelectromechanical systems (NEMS).

Nanolithography is a very active area of research in academia and in industry.

OPTICAL LITHOGRAPHY

Optical lithography, which has been the predominant patterning technique since the advent of the semiconductor age, is capable of producing sub-100-nm patterns with the use of very short wavelengths (currently 193 nm). Optical lithography will require the use of liquid immersion and a host of resolution-enhancement technologies like phase-shift masks (PSM) and optical proximity correction (OPC) at the 32 nm node. Traditional optical lithography techniques below 22 nm are not cost-effective. Next-generation lithography (NGL) technique may be the answer below that point.

Other nanolithography techniques include:

1. X-ray lithography can be extended to an optical resolution of 15 nm by using the short wavelengths of 1 nm for the illumination. This is implemented by the proximity printing approach. This method is simple because it does not require lenses.

2. A method of pitch resolution enhancement which is gaining acceptance is double patterning. This technique increases the feature density by printing new features in between pre-printed features on the same layer. It can be adapted for any exposure or patterning technique. The feature size is reduced by non-lithographic techniques such as etching or sidewall spacers.

3. Optical maskless lithography tool uses a digital micro-mirror array to directly manipulate the reflected light without the need for an intervening mask. Throughput is inherently low.

4. Electron-Beam Direct-Write Lithography (EBDW), the most common nanolithographic technique, uses a beam of electrons to produce a pattern of polymeric resist such as polymethyl methacrylate (PMMA).

5. Extreme ultraviolet lithography (EUV) is a form of optical lithography using ultrashort wavelengths (13.5 nm). It is the most popularly considered NGL technique.

6. Charged-particle lithography, such as ion- or electron-projection lithographies (PREVAIL, SCALPEL, LEEPL), are also capable of very-high-resolution patterning.

7. Nanoimprint lithography (NIL) and its variants, such as Step-and-flash imprint lithography, LISA and LADI are promising nanopattern replication technologies. This technique can be combined with contact printing.

8. Scanning probe lithography (SPL) is a promising tool for patterning at the deep nanometre-scale. For example, individual atoms may be manipulated using the tip of a scanning tunnelling microscope (STM). Dip-pen nanolithography (DPN) is the first commercially available SPL technology based on atomic force microscopy.

9. The furthest developed NGL remains X-ray lithography which is extensible to 15 nm resolution by use of "demagnification" in the near field.

10. Atomic force microscopic nanolithography (AFM) is a chemo-mechanical surface patterning technique that uses an atomic force microscope.

BOTTOM-UP METHODS

Nanosphere lithography uses self-assembled monolayers of spheres (typically made of polystyrene) as evaporation masks. This method has been used to fabricate the arrays of gold nanodots with precisely controlled spacings.

It is possible that molecular self-assembly methods will take over as the primary nanolithography approach, due to ever-increasing complexity of the top-down approaches listed above. Self-assembly of dense lines less than 20 nm wide in large pre-patterned trenches has been demonstrated. The degree of dimension and orientation control as well as prevention of lamella merging still need to be addressed for this to be an effective patterning technique. The important issue of line edge roughness is also highlighted by this technique.

PRODUCTIVE NANOSYSTEMS

Productive nanosystems are the functional nanometre-scale systems that make atomically specified structures and devices under programmatic control. Manufacturing is performed to atomic precision.

Present-day technologies have their own limitations. Large atomically precise structures exist, in the form of crystals. Complex 3D structures exist in the form of polymers such as DNA and proteins. It is also possible to build very small atomically precise structures using scanning probe microscopy to manipulate individual atoms or small groups of atoms. Larger, more complex systems are not possible to build by combining components in a systematic way.

Principles of physics and examples from nature both suggest that it will be possible to extend atomically precise fabrication to more complex products of larger size, involving a wider range of materials. An example of progress in this direction would be Christian Schafmeister's work on bis-peptides.

Nanorobotics

Nanorobotics is defined as the technology of creating machines or robots at or close to the scale of nanometres (10^{-9} metres). Nanorobotics can be said to be the hypothetical nanotechnology engineering discipline of designing and building nanorobots. Nanorobots would be typically devices ranging in size from 0.1–10 micrometres and arc constructed of nanoscale or molecular components. Artificial non-biological nanorobots have not been created till date and it is a hypothetical concept only.

A nanorobot can be defined as a robot which allows precision interactions with nanoscale objects or which can manipulate with nanoscale resolution. Following this definition even a large apparatus such as an atomic force microscope can be considered as a nanorobotic instrument when configured to perform nanomanipulation. Also, macroscal robots or microrobots which can move with nanoscale precision can also be considered as nanorobots.

ENDNOTES

1. A rotaxane is a mechanically interlocked molecular architecture consisting of a "dumb-bell-shaped molecule" which is threaded through a "macrocycle" (see graphical representation). The name is derived from the Latin for wheel (rota) and axle (axis). The two components of a rotaxane are kinetically trapped since the ends of the dumbbell (often called stoppers) are larger than the internal diameter of the ring and prevent disassociation (unthreading) of the components since this would require significant distortion of the covalent bonds.

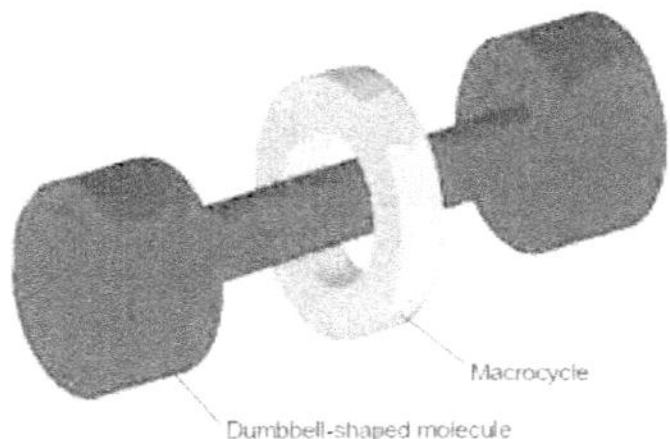

Graphical representation of a rotaxane, useful as a molecular switch

2. A catenane is a mechanically-interlocked molecular architecture consisting of two or more interlocked macrocycles. The interlocked rings cannot be separated without breaking the covalent bonds of the macrocycles. Catenane is derived from the Latin catena meaning "chain". They are conceptually related to other mechanically-interlocked molecular architectures, such as rotaxanes, molecular knots or molecular Borromean rings. Recently, the terminology mechanical bond" has been coined that describes the connection between the macrocycles of a catenane.

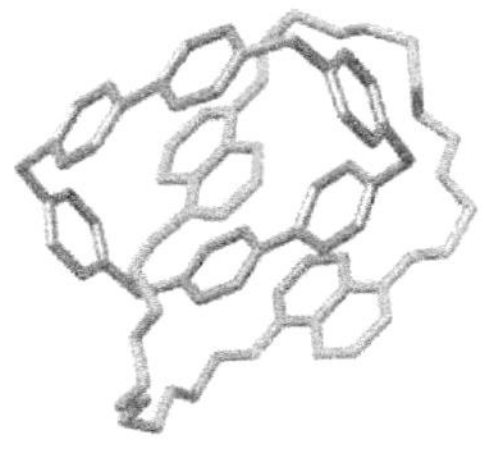

Crystal structure of a catenane with a cyclobis(paraquat-p-phenylene) macrocycle reported by Stoddart and co-workers in the Chem. Commun., 1991, 634–639

3. In supramolecular chemistry, an aromatic interaction (or π–π interaction) is a noncovalent interaction between organic compounds containing aromatic moieties.

4. Ultrahigh density, high-data-rate NEMS-based AFM data storage system. IBM Research Division, Zurich Research Laboratory, 8803 Ruschlikon, Switzerland. P. Vettiger, J. Brugger, M. Despont, U. Drechsler, U. Durig, W. Haberle, M. Lutwyche, H. Rothuizen, R. Stutz, R. Widmer and G. Binnig.

5. Nanotribology is a branch of tribology which studies friction phenomenon at the nanometer scale. The distinction between nanotribology and tribology is primarily due to the involvement of atomic forces in the determination of the final behaviour of the system.

6. Configuration entropy is the portion of a systems's entropy (a turning towards) tht is related to the positions of its constitutent particles rather than to their velocity or momentum.

REVIEW QUESTIONS

1. Discuss in detail about molecular machines.

2. What are the types of molecular machines?

3. What is nanolithography?

4. Write short notes on:
   i. Molecular machines
   ii. Synthetic molecular machines
   iii. Biological molecular macines
   iv. Molecular wires
   v. Nanomechanics
   vi. Optical nanolithography
   vii. Productive nanosystems

REFERENCES

1. Ballardini, R., Balzani, V., Credi A., Gandolfi, M.T. and Venturi, M. (2001). "Artificial Molecular-Level Machines: Which Energy To Make Them Work?". *Acc. Chem. Res.* 34 (6): 445–455.

2. Bhushan, B. (ed.). (2007). Springer Handbook of Nanotechnology, 2nd edition. Springer.

3. Despotuli, A.L. and Nikolaichic, V.I. (1993). "A step towards nanoionics." *Solid State Ionics.*60: 275–278.

4. Dhara Parikh, Barry Craver, Hatem, N. Nounu, Fu-On Fong and John, C. Wolfe. (2008). "Nanoscale pattern definition on nonplanar surfaces using ion beam proximity lithography and conformal plasma-deposited resist." *Journal of Microelectromechanical Systems.* 17: 3.

5. James, E. Hughes Jr., Massimiliano Di Ventra and Stephane Evoy. (2004). *Introduction to Nanoscale Science and Technology (Nanostructure Science and Technology).* Springer, Berlin.

Chapter 6
Molecular Manufacturing

DEFINITION OF CONCEPT AND HISTORY

Molecular nanotechnology, also called molecular manufacturing (MM) is the concept of designing engineered nanosystems or nanoscale machines operating on a molecular scale. It is related to the concept of a molecular assembler, a machine that can produce a desired structure or device, atom by atom using the principles of mechanosynthesis. Manufacturing in the context of productive nanosystems is different from the conventional technologies used to manufacture nanomaterials such as carbon nanotubes and nanoparticles. In simple terms, molecular manufacturing deals with production of machines which are meant for production of nanomaterials and does not deal with production of nanomaterials directly. Eric Drexler explained nanotechnology as nothing but future manufacturing technology based on molecular machine systems.

It is predicted that developments in nanotechnology will soon make possible the molecular machine construction by using biomimetic principles. Initially to be implemented by biomimetic means, ultimately advanced nanotechnology could be based on mechanical engineering principles—this is the projection of Drexler. The proposed line of development in manufacturing technology is based on the mechanical functions of the components like gears, bearings, motors and structural members. These components would be of help in programmable, positional assembly to atomic specifications. The physics and engineering performance of exemplar designs are described in Drexler's book *Nanosystems*.

To assemble devices on the atomic scale is very difficult as the process involves positioning atoms over other atoms of comparable size and stickiness. Carlo Montemagno was of the view that future nanosystems will be hybrid of silicon technology and biological molecular machines. Richard Smalley's view was that mechanosynthesis is impossible due to the difficulties in manipulating individual molecules echanically.

Carlo Montemagno   Richard Errett Smalley

Molecular-scale biological analogies of traditional machine components have demonstrated that molecular machines are possible theoretically. Non-biological molecular machines are at the initial stage of development. Three distinct molecular devices whose motion is controlled with the help of computers by changing voltage were developed, namely:

1. Nanotube nanomotor

2. Molecular actuator

3. Nanoelectromechanical relaxation oscillator

Like these devices, positional molecular assembly is a methodology initially developed in molecular nanotechnology. It was done using a scanning tunnelling microscope, to move an individual carbon monoxide molecule (CO) to an individual iron atom (Fe) sitting on a flat silver crystal and bound them chemically by applying a voltage. It was the beginning of molecular nanotechnology methodology and techniques.

So, molecular manufacturing means the use of precise, engineered, computer-controlled, nanoscale tools to construct many improved tools as well as products with precise, engineered nanoscale features. For this, a number of design principles and useful techniques for molecular construction via nanoscale machines have to be developed.

There are two types of approaches to achieve this target. One approach is from basic tools to integrated nanofactories. The next approach is, from integrated nanofactories creating an advanced nanofactory.

The differentiating points regarding molecular manufacturing and nanomanufacturing are:

1. Nanomanufacturing refers to advances in processing conventional materials, using large processing equipment and nanoscale phenomena to make small, usually simple products, such as nanoparticles.

2. Molecular manufacturing is a fundamentally different approach, using nanoscale machines to construct engineered, heterogeneous, atomically

precise structures by direct manipulation. Its ultimate goal is to build complex products, both small and large, with atomic precision.

APPROACHES TO THE STUDY OF MM

Molecular manufacturing will lead to the development of extremely high-performance nanoscale machines, far exceeding anything available today in engineering or in biology according to simple measures of performance, is the expectation. These machines can be integrated into manufacturing systems of any desired scale, capable of processing their own mass in hours or minutes. Products built with these manufacturing systems would be extremely valuable.

Molecular manufacturing can be studied using the following approaches.

1. Goals of molecular manufacturing or targeted development of MM
2. Basic theory of molecular manufacturing
3. Primary (basic) molecular manufacturing systems
4. Development of basic MM system
5. High performance nano- and microsystems
6. Nanofactory
7. Advanced nanofactory architecture
8. Mechanosynthesis
9. Molecular manufacturing products—advanced product design and performance
10. Research and development

GOALS OF MOLECULAR MANUFACTURING

The goals of molecular manufacturing or targeted planning of molecular manufacturing can be achieved only by long-term planning. The theoretical concepts would be practically possible in due course of time.

The high performance expected from the point of view of research and industry is responsible for, and forms the stimulus for, the fast development of MM. Factors for stimulus and planning for a targeted development program are to be made out.

Capabilities—Available and Expected

In theory, an advanced nanofactory-type molecular manufacturing system should have the ability to build precise structures of conductors and insulators. The conductors can be used for switches, as well as electrostatic motors and solenoids[1] and that is sufficient to build digital logic and actuated machinery.

In practice, the nanoscale technology is doing lot of new things. Recently, non-bleaching chromophores, mechanical programmable DNA-building systems made of DNA, several new kinds of molecular actuators and many relevant technologies have been developed. This derives future directions from the existing nanoscale technology research.

In conventional nanoscale technology, development of nanoscale-construction equipment is not the goal. Nanoscale mechanosynthesis helps in the synthesis of stiff covalent solids. Soft and flexible molecules from solution chemistry and biomolecular research are not suitable for conventional mechanical engineering at the nanoscale. Stiff covalent solids with lot of modifications from macro-scale practice may be suitable.

The emphasis on mechanical rather than electronic functionality has limitations but has advantages. Nanoscale mechanics depends on atoms, which are far heavier than electrons. The basic task is to join molecules repetitively and reliably, in programmable location or sequence and then to perform ordinary mechanical operations on the finished product available.

In short, a targeted molecular manufacturing development program would encourage basic research in a wide range of nanoscale technologies. This will occur during development of nanoscale substances, tools and engineering disciplines.

Evaluation

Maximum capability of molecular manufacturing and its products are to be evaluated. The next evaluation is related to time frame of the achievements and comparison with the existing technologies. Development of primitive general-purpose nanoscale manufacturing at the earliest would itself facilitate rapid development of improved versions. The physical limits of the products of molecular manufacturing need not deter an accelerated molecular manufacturing development effort. This will access the capabilities in advance of expected schedule of development.

Cost reduction of products is a possibility. Computers for simulation are becoming more capable, along with core molecular biology technologies that may be useful in development. Molecular manufacturing theory is continually advancing, finding cheaper and faster development pathways.

Guidelines

A program targeted at rapid development of advanced molecular manufacturing should have goals of futuristic ways. There are at least three possible pathways to develop an advanced system.

1.   Direct scanning-probe manufacture of stiff machines

2.   Engineering of biomolecules to develop an improving series of biomimetic machines which in turn will lead to high-performance integrated systems

3.   Fabrication of small molecular building blocks to build simple and improved nanomachine block-handling systems.

Simulation of the expected goals or ideas will lead to development of tentative designs in advance of laboratory capabilities. If the simulations are accurate, development of workable designs can help to create a further step and a stockpile of designs verified in simulation will be of use in speeding development. Every idea needs to be critically evaluated before effort is taken on the same. Efforts should aim at achievements, and the ideas are to be dealt in single, viable or multiple ways.

Modular design, in which the goal is broken down into sub-goals which can be solved in many ways, will make it easier to plan multiple parallel efforts to solve the problem. For example, the method of joining building blocks has some effect on the design of the actuators. A well-designed and broadly conceived specification for actuator characteristics will make a range of compatible block and actuator technologies to be developed in parallel.

BASIC THEORY OF MOLECULAR MANUFACTURING

Molecular manufacturing (MM) depends on the development of machines and the physical, chemical and mechanical aspects of workability at the nanoscale level. The laws governing the functioning of such machines are theoretically worked out and analysed. These laws are enumerated in an easily understandable way for a better understanding of molecular manufacturing.

Basic concepts of MM include factors related to nanosystems, including nanoscale manufacturing systems, which are discussed in detail.

Laws of Nanoscale (Scaling Laws)

The different parameters of performance of a machine improve as the machine size is reduced. Machines can be characterized and categorized by simple measures and ratios. A manufacturing system can handle its own mass of materials in a certain number seconds and a motor will handle a certain amount of power per unit volume. These numbers vary according to the size of the system. These relationships are called "scaling laws." or "laws of nanoscale". The relationship of a number of parameters with the functioning of the machines are detailed.

1. *The speed of linear motion is constant irrespective of scale.* When a machine is miniaturized, its frequency of operation will increase proportionally, since motion of machines cross less distance due to the minimized size and taking less time due to same speed. Because of this, the relative output of a manufacturing system, namely,

the fraction of its own mass that it can process per second will increase proportionally, assuming the components it handles scale down along with the machine. Each motion transfers a part proportional to the size of the device, but at a higher frequency.

2. *Mass of a machine decreases as the cube of the size of the machine.* Because of the mass and size relationship, the relative throughput of a machine is reduced by a factor of 10 and this reduction will additionally increase by a factor of 10, considering the cube factor. So, its absolute throughput will decrease by a factor of 100 (10 × 10). To maintain throughput as machines are minimized to the nanoscale, vast number of machines are needed. The rapid decrease in mass with size denotes that gravity will be unnoticeable in nanoscale machinery, momentum and inertia will be extremely small and acceleration will be correspondingly high.

3. *The linear relationship between size and relative throughput means that the machine handles components scaled to the size of the machine.* If the component size is held invariant (e.g. small molecules) as the machine scales, then the dependence of relative throughput on machine size is far stronger. For example, a 10-cm machine such as a scanning probe microscope would take on the rough order of 10^{18} years to manipulate its own mass of individual molecules. But the cube scaling of mass, combined with the linear scaling of operation frequency means that a 100-nm machine could manipulate its own mass of molecules in 30 seconds.

4. *Power density varies inversely with machine size.* This is explained by the fact that forces are proportional to area and decrease as the square of the size, while volume is proportional to the cube of the size. Speed being constant, power is force × speed. Power density is power divided by volume. This implies that nanoscale motors could have power densities on the order of terawatts (10^{12} watts) per cubic metre. That means, it would be difficult to cool large aggregates of such motors.

5. *In systems with property of wear, the lifetime decreases proportionally with the size.* This has been a serious problem for MEMS. But, due to atomic granularity and the properties of certain interfaces, atomically precise molecular manufacturing systems need not be subjected to incremental wear. Just as replacing an analog system with a digital one replaces analog noise with digital error probabilities, incremental wear is replaced by speed-dependent damage probabilities.

6. *Stiffness decreases in proportion to size, for a given material.* This will create design challenges for mechanical systems, but the use of stiff covalent solids will help. Diamond has a Young's modulus of about 1000 GPa and will be useful in such situations.

Effects of Atomic Granularity

At the energy levels at which nanoscale machinery operates, atoms are indivisible and all atoms of an isotope are identical. Two components built of atoms covalently bonded in the same arrangement will have identical shapes. Construction operations that are sufficiently precise to cause atoms or molecules to attach in predictable configurations can make perfectly precise products.

No post-processing steps to improve precision are needed. Machines of adequate reliability and repeatability can make perfect products. Product that is broken due to incorrect molecular structure is possible. If the products include manufacturing systems, then multiple generations with no loss of precision can be made.

Products built near atomic scale will have no wear, but can have breakage. The forces exerted by nanoscale machinery will be too small to break inter-atomic bonds. The granularity of atoms makes perfectly smooth surfaces impossible. But, smoothly moving bearings are possible.

Due to atomic granularity, machines cannot be designed with the dimensional precision common in macro-scale machining. The fact that atoms are soft and have smooth interactions reduces the impact of this limitation.

Information Delivery in MM

To create intricate, precise, heterogeneous nanostructures with individually specified features, one needs information. Enormous amount of information must be delivered from the computer, through the tools used, to the nanoscale.

A limitation of self-assembly is that the information must be embodied in the component molecules. An example is DNA structures that have been built containing thousands of bases. The synthesis process of such structures requires many thousands of seconds, corresponding to an information delivery rate of a few bytes per second.

Different methods of information deliveries used in nanotechnology are given as examples.

1. Scanning probe microscopes can be atomically precise; but, because their actuation systems are large, they move relatively slowly.

2. Semiconductor lithography masks require a long time to be manufactured, in spite of the fact that they can deliver information quickly. Once they are created, the overall data transfer rate from computer to nanoscale is low and the products are not precise.

3. Electron and ion beams may perform operations corresponding to kilobytes or even megabytes per second of information, but they are not atomically precise tools.

4. The scaling of operation speed indicates that to embody information in the manufactured product via rapid physical manipulation, it is ideal to use small

actuators. Nanoscale actuators, being smaller, will be able to operate faster and handle higher data rates.

i. Inkjet printers are an example as their print head actuators are a few microns in size and can deliver megabytes per second. Additionally, an inkjet printer can print its weight in ink in about a day.

ii. IBM's Millipede, a MEMS-based highly parallel scanning probe microscope array, can modify a substrate rapidly and is useful for computer data storage.

Both examples can produce two-dimensional "product" only, but inkjet technology has been modified to have three-dimensional products. Scanning probe arrays also have been used for dip-pen nanolithography (DPN).

To deliver a high-rate information stream to multiple actuators, to correct errors locally, and to use data formats, both actuators as well as digital logic should be at the nanoscale. Carbon buckytube transistors, single atom cobalt-based transistors, "crossbar," and the rotaxane switches are examples. Transistors can be built from individual molecules and that logic circuits can be built from supramolecular structures are also visualized.

Errors of MM

A non-zero error rate is the aim, but the rate can be quite low in well-characterized systems. Wear and manufacturing waste will not be sources of dimensional variation in covalent components. The major source of dimensional variation will be thermal motion, significant in nanoscale systems. Quantum and Heisenberg uncertainty[2] are having much smaller effects.

1. Thermal motion can cause positional variance and error. The effect of thermal motion on position can be estimated from the stiffness of the component. To resist thermal motion driving a system from a set state to an undesired state, an energy barrier is needed.

2. Because of adverse scaling of stiffness with size, operations that require fractional-nanometer precision require careful design to provide adequate stiffness and alignment. Design and analysis indicate that 100-nm-scale machines built of stiff covalent solids will be stiff enough to maintain the position of an articulated component within a fraction of an atomic diameter.

3. Ionizing radiation is a source of damage and error. An ionizing event can disrupt the structure of a molecular machine. Non-ionizing photons may excite bonds and cause damage, but can be excluded by a fraction of a micron of metal shielding.

Error Handling of MM

In a kilogram of nanomachines, the number of machines will be large, but the percentage of damaged machines will be very small. The appropriate method of error handling depends on the architecture of the system. Error handling should be tailor-made to suit the kind of manufacturing system and also the kind of product. Some ways to increase production are:

1. Using many small independent fabricators, each capable of producing more fabricators or small products. If each fabricator is independent of the others, then failure of a fabricator can be ignored, since the effect on the number of products will be less and ignorable.

2. Using large integrated manufacturing systems. In such a system, failures must be detected. Any large nanomachine-based product, including integrated manufacturing systems, must expect and cope with a non-zero component failure rate.

Space or high-altitude applications can have complications. Cosmic rays not only increase the dose of radiation but also create a different pattern of damage. A cosmic ray is a very heavily charged particle, the nucleus of an atom, moving at very high speed. It does not create randomly spaced points of damage but creates a dense track. Where cosmic rays are present, error handling designs must cope with multiple simultaneous failures in adjacent machinery.

Scaling up Production in Scaled Down Machines

To make a gram of product, vast numbers of machines must be run in parallel in MM. Methods to achieve this involves manufacturing systems that build other manufacturing systems. As an example, about 60 doublings are required for one 100-nm machine to produce a kilogram of machines. There are two methods to achieve this.

1. One method is for small machines to produce many small products. By this, the mass of machines increases, but the machine and product size remain at the nanoscale. Large numbers of nanoscale products can be useful as networked computers, as pharmaceuticals or as components in larger products. Modular robotics may be used to make large products from aggregates of nanoscale robots.

2. Another method is to build integrated systems containing numerous manufacturing subsystems attached to a framework. This allows nanoscale building blocks to be joined together, forming a large integrated product. An attractive architecture for a large, advanced, integrated factory is a planar arrangement that produces blocks of micron size from each subsystem, attaching them to a growing product. A planar architecture is good for

primitive manufacturing systems, since each subsystem deposits its portion of the product on adjacent sites of a nearby planar substrate.

Small manufacturing systems require less internal architecture than integrated systems, but may be difficult to interface with external controllers. It would be difficult to supply high-bandwidth information and power efficiently to free-floating machines. Small products can be useful but large integrated products have a much broader range of applications.

Energy Requirements

The large number of operations necessary to build large objects by manipulation of individual molecules require more energy. A covalent reaction may dissipate around 500 zJ, although this energy can be recovered if the reactants are held by mechanisms in a sufficiently stiff manner to avoid "snapping" from one position to another.

Design

It is often useful for a subsystem to maintain a certain state until pushed to the next state by another subsystem. This is useful for at least two reasons. It simplifies the control architecture and it saves the energy that would be required to "compress" a system from an unknown state into a known state.

An efficient design will attach to a system in a known state, remove the barrier, move the system to the new state, replace the barrier with the system now on the other side, and then release the attachment. The only losses in such a mechanism will be frictional.

PRIMARY (BASIC) MOLECULAR MANUFACTURING SYSTEMS

The primary or basic or primitive manufacturing system development is in theoretical proposition and is to be worked out practically.

A general-purpose manufacturing system capable of building duplicate and improved systems is the agenda. Such a system is yet to be developed. The theoretical possibilities using today's technology identify two architectures for primitive systems to be developed. Both these theoretical models use relatively uncomplicated nanomechanical systems, attached to larger actuators, to bind prefabricated molecular "nanoblocks" from solution and fasten them to chosen locations on the product. Each design is planned to be capable of manufacturing an array of independently controlled duplicate systems.

History and Background

The first molecular manufacturing systems will be based on solvent-immersed mechanisms assembling prefabricated molecular building blocks or on scanning probe systems doing machine-phase mechanosynthesis to build covalent solids. Recent progress in molecular building blocks denotes that both primitive "wet" system and advanced "dry" system are the possibilities.Deriving examples from biology, different classes of molecules can implement an engineered structure like that of DNA. Nadrian Seeman has made crystals up to a millimetre in size, ordered to 1 nm resolution. Protein design has resulted in novel folds and can be used to produce small complex shapes. Christian Schafmeister has developed a polymer that is stiff even in singlestrands, using spiro linkages between cyclic monomers.

To form a solid product, blocks must fit together. To help with alignment and insertion, a completed layer of blocks should form shallow pits between the blocks into which the next layer of blocks will fit. Each completed layer forms a regular grid of vacancies for he next layer of blocks to be placed into (Figure 6.1).

Figure 6.1   Depositing nanoblocks via scanning probe to build an array of "Tattoo" block-placement machines

Tattoo Type of Architecture

The Tattoo architecture for programmable heterogeneous assembly of nanoblocks is based on the possibility of making nanoblocks that will not aggregate or bond in solution, but will bond when pushed together by mechanical force.

Once the ability to build patterns of blocks is established, the next step is to build a "tattoo needle." This is a socket attached to a nanoscale actuator[3] which can be individually activated via electrical connection, if the actuator is electrical.

Construction of the socket will probably require special design (Figure 6.2). Several blocks placed in a triangle will make a block-shaped cavity. Post-processing step by chemical means maybe required to modify the special surfaces of the blocks.

Figure 6.2   Two ways to implement a socket: (a) Place several blocks to form a cavity, perhaps followed by post-processing to modify the inner faces. (b) Use a special-purpose molecular building block.

Once the actuator and socket machine is built, it can be used to deposit blocks on a surface held in front of it. The machine should be able to build a second, identical machine on the surface that is presented to it by the manipulator and a row or array of machines. These machines could then be used in parallel for higher-throughput manufacturing of larger arrays.

If the machines can be independently controlled, they will be useful to build heterogeneous structures or regular structures. If different types of nanoblock-specific sockets can be built on different machines in the grid, then multiple nanoblocks can be mixed in one feedstock solution and each machine dealing with a particular block is activated, only when its selected type of nanoblock is wanted.

Simple approach might be useful to test the block-binding surface functionality before complicated nanoblocks and nanoscale machines like towers are developed.

Silkscreen Type of Architecture

The silkscreen architecture is based on the concept of separating a solution containing nanoblocks from the substance or condition that would cause the blocks to bond together. The silkscreen is a membrane with an array of holes. It separates the feedstock blocks from the product and can maintain distinct conditions like concentrations of zinc on each side. Its primary purpose is to control the position and timing of block passage through the membrane to the product.

Each hole in the membrane contains an actuator which can reversibly bind to a single block, transport it through the membrane, block the hole to prevent mixing of solutions and present the block to the product.

The membrane would be closely fitted to the growing product and could be aligned to it by local forces. A block passing through a hole in the membrane must be able to reach only one vacancy in the product. The block's motion can be constrained by the manipulator until it is bound to the product.

An initial silkscreen membrane might be built by any convenient combination of self-assembly, synthetic chemistry and lithography. The grid of holes could be created either by lithography or by a self-assembled membrane such as DNA tiles. Each hole would be filled by a molecular actuator system. Once constructed, the first system could be used to build larger membranes and improved designs.

The simplest membrane might have only one hole. Its actuator could be activated by light or by passing current between the product and feedstock side of the membrane. The hole might be constructed by slow lithographic techniques such as ion milling or dip-pen nanolithography. Even with only one hole, the size of the product would be limited only by the speed and reliability of deposition and by the range of the product positioner. The steps involvd in silkscreen type of architecture are shown in Figure 6.3.

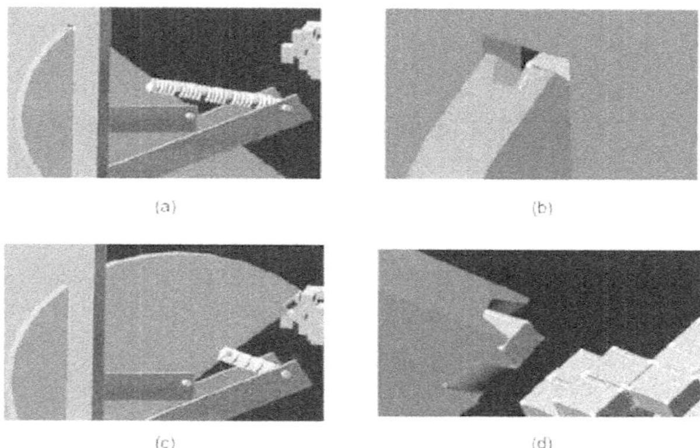

Figure 6.3   (a) The manipulator is in position to gather a block from solution. (b) A close-up of the socket's position with a block inserted. (c) The manipulator has moved to attach the block to the product. (d) A close-up of the block about to be pressed into place.

Nanoblocks

The feedstock of a primitive "wet" molecular manufacturing system will be prefabricated molecular or macromolecular blocks, a few nanometres in size, dissolved or suspended in a solvent. The function of the manufacturing mechanism will be to take individual blocks from solution, move them to a specified location in the product and attach them strongly enough to bind. The design of the system must ensure that blocks in solution do not aggregate with each other or attach to the product where they are not supposed to, but once attached and bonded to the product they should not detach.

There are many ways in which blocks can be strongly joined which include a wide range of chemical reactions or usage of intermolecular forces.

For the purpose of joining blocks, the reaction should happen only when the blocks are pushed together, not when they are floating in the feedstock solution. If zinc binding is used, then excluding zinc from the feedstock solution and excluding stray blocks from the product area can prevent the feedstock blocks from aggregating in solution. The "silkscreen" approach was designed to maintain a product environment of different composition than the feedstock solution.

Attachment mediated by photochemistry or electricity should work with either the tattoo or the silkscreen approach.

In order to manipulate blocks mechanically, they must be attracted from solution and attached to a manipulator. This will happen if a binding site ("socket") in the manipulator is made attractive to the blocks, for example by giving its interior a charge opposite to the charge on the blocks. This is an application of self-assembly.

The manufacturing system will need to be able to place more than one kind of block. There are two ways to accomplish this. One possibility is that the block types will be mixed in the feedstock solution, in which case the sockets must be block-specific. The other possibility is to flush through one feedstock solution at a time, with each solution containing a different block type. When a solution is flushed out, blocks will remain in sockets, but can be dealt with simply by depositing them onto the product in an appropriate location.

If the system uses zinc binding, then putting a few zinc binding sites in the binding surface of the socket can be used to bind to the block strongly enough to hold it reliably, but weakly enough to let it go without damage when the actuator is retracted from the product.

Nanosystem Functions

The manufacturing system of simple type requires to extend or retract a bound block. This in turn needs a linear actuator with one degree of freedom and a small range of motion and a method to control individual actuators.

Speed and addressability will be important for any practical nanofactory. Molecular precision and small size are important.

Some molecules are responsive to light. Light has the advantage that it requires no physical structure to deliver it. It can be focused in a pattern from a distance. Light can be switched with very high bandwidth. A focused pattern of light will have low spatial precision and may be partially overcome if molecules can be found that are sensitive to specific wavelengths of light.

There are several kinds of electrically actuated materials. Piezoelectric materials deform because the spacing between charged atoms in the crystal varies slightly under an electric field. Electrically deformable polymers working by ion intercalation/expulsion, redox reactions causing changes in a molecule's electron distribution are to be taken care of.

Sensing will be important to provide early confirmation that new designs are working as intended. The block can be detected by contacting it with a physical probe. Full extension of the probe would indicate an absent block.

Information must be returned from the nanoscale either optically or by electrical signal return using a simple mechanical switch and by fluorescent nanoparticles.

Throughput and Scale up of Nanoblock Placement

Activating a molecular actuator via photons or electric fields might take on the order of a millisecond. Diffusion of nanoblocks to fill the sockets might take several milliseconds. This depends on many factors, including block size, the concentration of the blocks and the viscosity of the solvent. Diffusion and binding of blocks to sockets is probabilistic and without the ability to detect when a socket is filled, a relatively long time must be spent waiting until it is very likely to be filled. If this is the limiting factor in deposition speed, the ability to sense when a socket is filled would allow faster deposition and scaling up of nanoblock placement.

Development of Basic MM System

The main goal of development of a basic system is to build a better one. Development of basic manufacturing system towards self-contained kilogram manufacturing system is the aim of development. It might have practical applications. The ultimate goal of MM is an easy-to-use nanofactory which can build a complete duplicate nanofactory like that of the one originally available as well as a wide range of high-performance products, from inexpensive feedstock. This is beyond the capacity of basic nanofactory. But a set of adding capabilities to basic unit can form a sequence of improvements ending at the goal of creating a better one. The capabilities could be added one at a time so that each capability can be developed and tested separately.

The capabilities are independent and to be added in any of the sequences of development. It is possible to validate future designs, to guide design choices and faster development.

Development Methodology

The following improvement planning may be useful.

1. Digital logic is necessary as the system scales up.
2. Molecular electronics with transistors made out of carbon nanotubes and single organic molecules will help. Simple electrical switches operated by molecular actuators may also be used.
3. With more efficient and intricate mechanical systems, more flexible and functional manufacturing processes without sacrificing efficiency are possible. Actuators and bearings can be improved at any stage.
4. Covalent bonds can be flexed without weakening and the range of motion is considerably higher than with MEMS.
5. Stiff smooth surfaces need new type of bearing.

Solvent Avoidance

There are many ways of fastening blocks into the product that do not require solvent. Pure electrostatic actuators are excellent choices for a dry environment and should also work in clean non-polar solvent.

A silkscreen type factory could maintain solvent on the feedstock side of the membrane and low-pressure gas on the other side. Liquid xenon may be a good choice for a solvent. A major advantage of xenon is that it is chemically inert, so it will not combine with any reactive molecules used in mechanosynthesis.

Many reactions require no solvent. Radicals that could not survive in solvent can be maintained in vacuum.

A dry nanofactory could work far faster than the wet versions, depending on the actuators. Electrostatic actuators should be very fast, limited only by the current carrying capacity of the wires.

Block Manufacturing

The nanofactory will be more efficient, its products will be cheaper and it will be easier to design and create new block types, if small molecules can be used as feedstock, combined into large blocks internally and the large blocks placed in the product layer. Small molecules have no room for elaborate molecular attachment mechanisms, but there are many approaches to joining molecules. There is a wide variety of ways to make programmed parts from molecular feedstock. Molecular fragments weakly bonded to "tool tip" molecules can be deposited on a depassivated surface in vacuum. This is called "machine-phase chemistry."

In solvent, molecules can be forced to bond to selected locations on a terminated surface. A chain polymer can be built which then folds up into a desired shape. Component molecules can be used which hold their position by electrostatic or van der Waals attraction or hydrogen-bonding.

Deposition reactions can happen through conventional, non-mechanical chemistry. It may be useful to build blocks out of molecules that are already structural or functional components.

Mechanical Design Improvement

Directed internal transport of product components within the factory would be useful for several reasons. It would allow broken machinery to be bypassed. When a factory that manufactures blocks internally is forming a product with large voids, internal transport would allow the transfer of blocks from the entire factory's block-manufacturing machinery to dense regions of the product, alleviating a potential bottleneck.

HIGH PERFORMANCE NANO AND MICRO SYSTEMS

Designs and techniques useful in advanced nanofactories and producs are discussed. It focuses on high performance nanoscale designs.

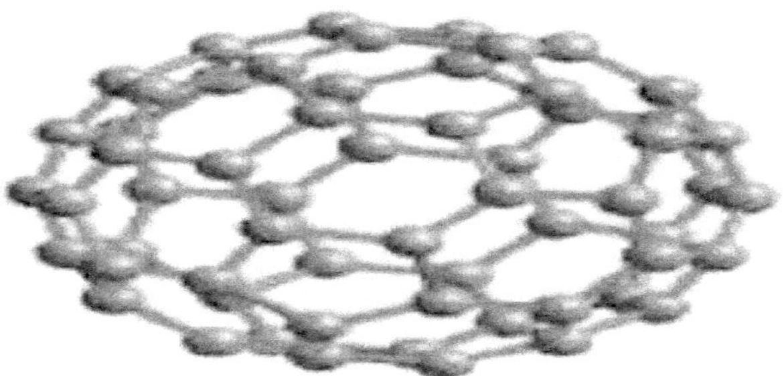

Figure 6.4   Bucky ball

Basic scaling analysis and theoretical designs using known materials make a projection that the final performance of nanomachinery may be higher than is achieved by either biological mechanisms or computers.

Diamond, graphene and fullerene can be produced by a wide range of reactions. New low-temperature records are continually being set for fullerene synthesis and solution-phase synthesis of C_{60} (buckyball) (Figure 6.4). The flexibility of "dry" chemistry appears well-suited to building diamond lattice by incremental deposition. A high performance nano-manufacturing technology uses materials of diamond-like strength.

Surface Forces and Component Manipulation

Due to electron interactions, an attractive force will develop between almost any object in close proximity. Although the force can be screened by liquids, in gas or vacuum it provides a convenient means for handling or attaching micron or sub-micron blocks.

Manipulation of micron-scale blocks does not require mechanical grippers. Contact pads of a few square nanometres area can implement the functionality of grippers.

If the manipulator is capable of bending motions, then rotating its pads on edge will allow them to disengage from the block, while a combination of translocation and slow bending will rotate the whole block and disengage it.

Mechanical Fastening: Ridge Joints

A strong mechanical joint (Figure 6.5) that requires low insertion force, but activates itself to lock in place can fasten micron-scale blocks together without complicated manipulations. Surface forces can be used to power te mechanism of the joint. "Expanding ridge joint" is an example.

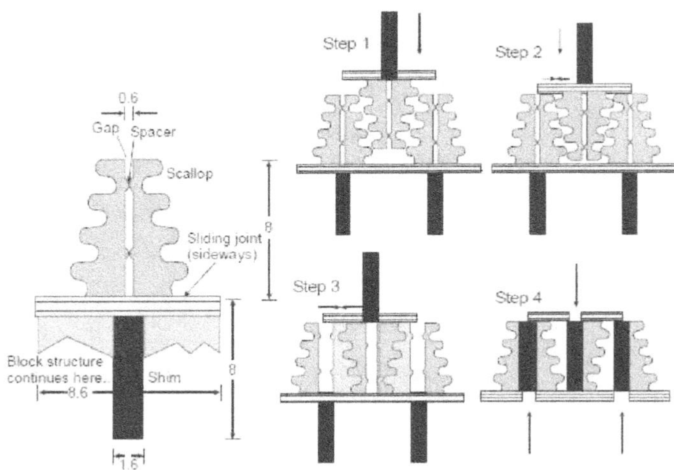

Figure 6.5 Operation of a ridge joint. Ridges are pulled apart by surface forces without external actuation or manipulation.

Bearings

The granularity of atoms cannot make a perfectly smooth surface but low-friction bearings can still be constructed out of stiff materials. Mechanically, atoms are soft, rather than hard-surfaced spheres. Bonds are also somewhat compliant. Atoms overlap when they bond, reducing the irregularity of the surface.

If the atoms on two facing surfaces have the same pattern and spacing, the atoms of one surface will fit between the atoms of the other, requiring high force to slide the surfaces. To prevent this, the atoms can be placed out of register, with different spacing or orientation. In this case, the transverse forces on the atoms will almost completely cancel, leaving a very small net force (and hence very low static friction) at any displacement. Super lubricity is a condition of very low friction between surfaces with atoms in non-corresponding positions and has been observed between graphite sheets rotated to out-of-register angles.

Nested carbon nanotubes (with different numbers of atoms on the inner and outer surface) have been observed sliding and rotating freely. This would allow building low-friction rotational bearings.

Atomically precise surfaces can slide past each other without friction or wear and they do not need lubricants. Without lubricants, the perpendicular stiffness of a sliding bearing is high, being a function of surface forces between two stiff surfaces.

Bearing using a covalent single bond is suitable for relatively low loads, but may be useful in some small machines. It is expected to have especially low drag.

Efficient nanoscale bearings are expected to have effectively zero static friction. The net force exerted by a bearing surface will usually be far smaller than the forces used to drive the machinery. The energy barriers created by bearing forces will be lower than thermal energy.

Electrostatic Actuators

At large scale, electrostatic actuators require high voltages and have low-power density. By running it in reverse, the same device would become a generator. Diamond is an excellent insulator. Some carbon nanotubes are excellent conductors. So, all carbon types will be useful in electricity usage of MM.

Digital Logic

A lower bound for the performance of digital logic can be set by a simple, easily analysed, purely mechanical design. Nanoscale rods that move to block the motion of other nanoscale rods can implement logic gates.

The mechanical approach to logic is relatively insensitive to material choice. Because it does not rely on electronic effects, components can be packed tightly without limits imposed by electron tunnelling. Error rates can be extremely low, because an error would require a logic rod to slip past a physical obstruction.

Mechanochemical Power Conversion

In reactions between small molecules or non-stiff components of large molecules, the bonding force drives the reaction to completion quite rapidly. However, if reactants are stiffly held, the reaction can be constrained to move slowly through the intermediate stages. This will exert a force on whatever is holding the reactants, and energy can be extracted from this force. It is estimated that the feasible power density of mechanochemical energy converters is on the rough order of 10^9 W/m^3.

NANOFACTORY

A nanofactory can be defined as a proposed system in which nanomachines like that of molecular assemblers or industrial robot arms would combine reactive molecules via mechanosynthesis to build larger atomically precise parts. Macroscopic or visible atomically precise products would be assembled by positioning mechanisms of assorted sizes, in turn.

A typical nanofactory would fit in a desktop box, in the vision of K. Eric Drexler. Various nanofactory designs are summarized in *Kinematic Self-Replicating Machines* (2004) by Robert Freitas and Ralph Merkle.

Personal Nanofactory

It is a proposed new appliance that might sit on a countertop in a home. To build a personal nanofactory (PN), there is need to start with a working fabricator, a nanoscale device that can combine individual molecules into useful shapes. A fabricator could build a very small nanofactory, which then could build another one twice as big and so on. Within a period of weeks, one can have a tabletop model. Products made by a PN will be assembled from nanoblocks, which will be fabricated within the nanofactory. Computer-aided design (CAD) programs will make it possible to create state-of-the-art products simply by specifying a pattern of predesigned nanoblocks.

Products of Nanofactory

1. Lifesaving medical robots.
2. Untraceable weapons of mass destruction.
3. Networked computers.
4. Networked cameras so that governments can watch everyone and every move.
5. Rapid invention of wonderful products for human use at unimaginable level.
6. Weapons development fast enough to destabilize any arms race.

Self-replication of Nanofactory

After constructing a single molecular assembler, it is programmed to self-replicate to produce many copies that will allow an exponential rate of production. After sufficient quantities of the molecular assemblers are available, they can be re-programmed for production of the desired product. If self-replication of molecular assemblers is not controlled, it may lead to ecophagy or the grey goo problem.

One method to building molecular assemblers is to mimic evolutionary processes employed by biological systems. Production of complex molecular assemblers might be evolved from simpler systems. Developing a system that works should be the goal and not simple architectural machines.

Most assembler designs should keep the "source code" external to the physical assembler. At each step of a manufacturing process, that step is read from an ordinary computer file and "broadcast" to all the assemblers. If any assembler gets out of range of that computer or when the link between that computer and the assemblers is broken or when that computer is unplugged, the assemblers stop replicating. This "broadcast architecture" is one of the safety features for grey goo.

Drexler and Smalley on Nanofactory

Drexler felt molecular assemblers would transform the world. They would build anything allowed by the laws that govern the universe with absolute precision and without pollution. Great potential dangers like the destruction of life on earth were also thought of. Such general assemblers are inevitable and we need to develop protocols to ensure their safe implementation.

Professor Richard Smalley who won the Nobel Prize for his contributions to the field of nanotechnology believed that such assemblers were not physically possible and introduced scientific objections to them. His two principal technical objections were termed the "fat fingers problem" and the "sticky fingers problem" which would exclude the possibility of precision picking and placing of individual atoms.

Regulation

Discussion of the possible implications of molecular assemblers created the necessity for regulation of current and future nanotechnology. There are concerns with the potential health and ecological impact of nanotechnology that is being integrated in manufactured products.

ADVANCED NANOFACTORY

Architecture and function of an advanced nanofactory is the next step of design in molecular manufacturing and is only theoretical in nature. The design is derived from:

1. Burch/Drexler planar assembly architecture

2. The Phoenix primitive nanofactory

The planar assembly structure suggested for the primitive molecular manufacturing system is also useful for high-performance nanofactories. The blocks that would be attached to the growing product would be larger than in the primitive design, of the order of 100 nm to 1 micron.

Blocks would be manufactured from small feedstock molecules and then transported to the assembly face. Maximum product deposition speed would be of the order of 1 mm/second, though the rate of deposition depends on the rate of fabrication of blocks unless prefabricated. Feedstock and cooling fluid would be distributed to the input side of the nanofactory via pipes. Design and analysis indicate that a 1-kilogram factory might manufacture and deposit 1 gram of product per second. In a kilogram of nanomachines, errors are inevitable, so each section of the factory must be designed for error detection and redundancy.

Raw Materials of Nanofactory

In an advanced nanofactory, the raw materials are blocks which need fabrication, delivery of blocks, and control of block delivery.

Block Fabrication

The most flexible way to build large (million- to billion-atom) blocks from molecules is to use a general-purpose manipulator system to add molecular fragments one at a time to the block. One way to increase the output of the process is by building several block components in parallel, either at general-purpose work stations or on special-purpose fixed-path mills and then combining them to form the block. The sub-parts could be either general-purpose parts, such as lonsdaleite cylinders to be added to a diamond crystal, or special-purpose parts like computational elements. Either way, this could speed up block construction several fold, allowing deposition to begin sooner.

The fastest way is to prefabricate them using any convenient combination of synthetic chemistry, self-assembly, mechanosynthesis and simple mechanical assembly.

Block Delivery

Product deposition speed will be limited by three factors namely:

1.   block delivery speed,

2.   block placement speed and

3.   block fabrication speed unless blocks are prefabricated.

A kilogram of blocks distributed over a square metre would make a 1-mm thick layer. The use of high-strength covalent solids makes the product to have very low density. Product structural members may have inflated volumes due to increased wall thickness or composed of light-weight truss or foam. Accordingly, it is necessary to provide a large product deposition area even for a kilogram-per-hour factory.

For efficient operation, the fabrication of blocks can continue during times when the mass deposition rate is lower than the mass fabrication rate. When high concentrations of mass are to be deposited, blocks can be delivered from the entire fabrication volume. In that case, blocks will have to be transported from the area where they are manufactured to the area where they are deposited.

Blocks will be large enough to be handled mechanically and can be transported by any convenient mechanism. A layer of small "cilia" that pass the block overhead is mechanically simple and allows redundant design. Grippers are unnecessary and contact points for surface forces are sufficient.

Depositing a thin, solid column of blocks at maximum speed requires blocks to be delivered from manufacturing sites from all over the factory to a small area of fabrication. This determines the thickness of the routing structure, which is roughly the same as the desired thickness of the deposited part divided by the ratio of delivery speed to deposition speed. For example, to allow deposition of a 1-cm solid block at 1 mm/s, if the blocks can be moved internally at 1 cm/s, then the routing structure must be 1 mm thick.

Nanofactory Architecture

Physical architecture, power supply, cooling of the system and control mechanisms of nanofactory are parts of nanofactory architecture.

Physical architecture   The product is deposited from a block-placement plane which is studded with manipulators to push the blocks to the product surface. The manipulators may hold the growing product, making it possible to build several disconnected parts that would be fastened together after further deposition. Alternatively, the product may be held externally. This would require disconnected parts to be connected by temporary scaffolding.

Between the manipulators are holes through which blocks are supplied from the routing structure. The routing structure is composed of braided or interwoven block delivery mechanisms (probably cilia). This allows blocks to be shipped crosswise, routed around damaged areas and so on. Power and control for the delivery mechanisms, as well as for the placement plane mechanisms, run inside the delivery structure.

Below the routing structure is the fabrication volume. This contains most of the mass of the nanofactory. For a kilogram-scale factory with a square-metre deposition area, the fabrication volume will be about 1 cm thick. This provides about 10,000 square metres of surface area for feedstock intake and cooling. Diffusion and heat transfer will not limit the nanofactory speed.

Because the physical architecture of the nanofactory is planar, with feedstock intake and processing located adjacent to product deposition, there is no need to change any dimension of the factory's nanoscale components in order to increase the manufacturing capacity and deposition area. In effect, multiple square-metre designs can be abutted to make as large a factory as desired.

Power and cooling   Most of the energy used by the factory will be used in the block-fabrication area, since handling the smallest components (feedstock molecules) will require the majority of operations. This is the area that is closest to the cooling/feedstock fluid. A cooling fluid made of a low-viscosity carrier of small encapsulated ice particles can provide very efficient heat removal. Cooling by phase change also has the advantage of keeping the whole factory at a uniform and constant temperature. Within the nanofactory, power can be distributed very efficiently at all scales by rotating shafts. Electrostatic motor/generators can be used to interface with an external electrical power system.

Factory control and data architecture   Any system capable of building a nanofactory should also include large numbers of digital computers to implement digital logic. General-purpose (microprocessor style) computation is energetically expensive, so it will be infeasible to use many CPU cycles per atom in a high-throughput kilogram-scale nanofactory. Instead, special-purpose logic (state machines) can be used to good effect for repetitive feedstock-placement tasks. To plan the handling of larger blocks (many millions of atoms), general-purpose computation will add minimal cost.

The control system will be deterministic and detailed, making it possible to specify any nanoblock at any position. One nanoblock design can be re-used at many locations. Arrangements of nanoblocks can also be copied (tiled) and used to fill solid geometry volumes.

There are only about 10^{15} 100-nm blocks per gram. With the computing resources in a kilogram-scale nanofactory, it is possible to plan the path of each individual block from where it is fabricated to where it will be placed.

The blueprint/control file will be broadcast equally to all parts of the factory. This allows a large number of local computers to be accessed with a simple network design.

The distribution planning is complicated since concentrations of mass in different parts of the product may draw blocks from all over the factory. Preprocessing along with tolerance for mild inefficiency will allow block production and delivery to be planned with simple deterministic algorithms.

Fabrication instructions for individual blocks will be delivered next. Each fabricator will know which part of the instruction stream to pay attention to. In practice, a local computer will likely control multiple fabricators, parsing the instruction stream and sending appropriate instructions to each fabricator it controls.

This plan assumes that all fabricators will be working in parallel. This can be accomplished by having broader communication channels to each local computer. Blocks will probably take minutes to hours to build, which is enough time to transmit many gigabytes of data.

Quality control

Environmental cleanliness   The internal environment of the factory must be kept clean from contaminant molecules that could cause undesired reactions or jam the moving parts. The factory interfaces with the environment at feedstock delivery and at product deposition. The feedstock molecule intake mechanism will deterministically manipulate the desired molecules, which provides an opportunity to exclude other molecules. This will be relatively easy if the molecule is small, like acetylene or at least compact so that a shaped shell closed around it can exclude all solvent molecules. Small feedstock is preferable for several other reasons as well, including feedstock cost and flexibility of manufacture.

Environmental contaminants can be kept out of the product deposition mechanism by extruding a thin shell or sealing sheet to cover the product and any unused area of the deposition surface. Before the product is removed, a second covering must be deposited to seal the deposition surface.

Reliability   Most of the mass of the nanofactory will be used for manufacturing blocks from feedstock. This implies that many fabricators per second will be damaged and a percentage of blocks under construction will not be completed. This can be dealt with by building duplicates of all block types.

Blocks damaged after fabrication while in transit to the product assembly surface need not be replaced. The product's design will need to cope with radiation damage immediately after manufacture and its lifetime will be far longer than the manufacture time.

The transport mechanism will consist of many redundant arms/struts/cilia that work together to move the block along surfaces. The random failure of a small percentage of these will not compromise the ability of the rest to transport the blocks. If a part of the transport mechanism fails, blocks can be diverted around the area.

There is limited room at the planar assembly surface and the robotics may be more complex than for block transport. However, the volume of radiation-sensitive machinery is correspondingly small. If repair occurs, entire placement operation could stop during each repair without greatly reducing performance. Another solution is to make the placement machinery flexible and large enough that machines can reach to do the work of a disabled neighbour.

MOLECULAR MANUFACTURING PRODUCTS

Molecular manufacturing products are expected to be of high order of performance. To get high performance of nanoscale machines and materials, it is to be planned that large products combine with large numbers of nanoscale components in efficient structures. Nanoscale performance advantages can be preserved in large-scale integrated systems is the observation by preliminary architectural work. Large aggregations of high-powered machinery need to be cooled. This cooling problem will produce less effect regarding performance than presently available products. The industry experts the materials and products to be of high performance and so, the design is to be made accordingly.

High-performance Materials and Properties

The properties of high performance materials in nanotechnology are:

1. *Resistant to flaws and failure*   Strong, stiff solids when put into tension by external forces are easily weakened by minor flaws. The flaws will lead to failure and in turn the failure will propagate rapidly. Solid carbon lattice (diamond) can behave like that. Crack propagation can be prevented. It can be done by building the material in long thin fibres attached in a way that does not propagate failure. Similar types of approaches are used today in advanced polymers and fibres, but molecular manufacturing construction would give more control over structure. This would allow fibres to be perfectly aligned and cross-linked to each other or attached to other structures with minimal strain. But the fibres should be defect-free.

2. *Energy absorption efficiency*   Strong fibres can also form the basis of energy-absorbing materials. Interlaced fibres with high-friction molecular surfaces could be designed to slide over each other under stress slightly less than that needed to break bonds in the material. The limit to the energy absorbed in such an "unbreakable" material is the heat capacity of the material. When its temperature rises too high, its bonds will be weakened. In theory, this much energy could be thermalized (absorbed) in just a few nanometres of motion. Longer fibres would allow the material to absorb repeated impacts after it had cooled.

3. *Solid cum space model*   A solid block, slab or beam of material typically is not efficient at resisting compressive stress. Variety of efficient structures like honeycombs and fractal trusses are the goal. A thin pressurized tank will resist compression at any point, transferring compressive stress to the contents and imposing tensile stress on the walls.

4. *Energy storage*   A diamond shaft rotating at high speed can carry power at 10 watts per square micron. This may be the most compact way to transmit power within a product. Stretching a spring made of tough diamond structure can store energy equal to a significant fraction of the bond energy of the spring's component atoms. Such a storage system could be charged and discharged quite rapidly, and store energy without leakage.

Advanced Products and Performance

The performance of a product will depend on the following parameters namely mass, power and heat budgets.

1. **Mass**   By adding more systems and running them more slowly to obtain reduced drag, mass can be programmed for efficiency.

2. **Power**   Taking into account the high-power densities of electrostatic motors and the smaller yet high-power density of electrochemical fuel cell or mechanochemical processors, power transformation will not be a significant part of the volume of most metre-scale products. If the product expends its power externally, like for propulsion through a viscous medium, a small fraction (within 1%) of the total power handled need be dissipated as internal frictional heating.

3. **Heat budgets**   Massively parallel computer systems can generate heat and are an important source of heat in a system. Increasing the size of the computer can increase the signal path length, requiring slower operation. A computer using reversible logic will be most efficient when it spends about half of its energy on erasing bits, and half on friction.

Computers will draw very little power presently. For many applications, new algorithms would be required to make use of such a highly parallel system. Cooling a cubic-centimetre volume of computers can be accomplished via branched tubing and a low-viscosity coolant fluid using suspended encapsulated ice particles.

Using the full strength of diamond, handling compressive stress efficiently and using much more compact motors, computers and sensors, products could be built with a small fraction—usually less than 1%—of the structural mass required.

Advanced Products and Designs

When product is expected to be complex, it is to be manufactured via relatively simple and crude processes and the number of operations is to be minimized to reduce manufacturing cost.

The goal of designing such complex products is to re-implement existing products in the new technology. Presently, a product in use is utilized for a single function namely a motor, a computer or a structural beam. Many of these components should be able to be replaced by a higher-performance component without changing the product's functionality. Replacement of electromagnetic motors by nanoscale electrostatic motors may require a flywheel to be added to replace the missing rotor mass.

For replacement of components of a product as available presently, the key ingredients are well-characterized nanomachines and design libraries that combine them into larger structures.

Design of nanofactory-built products may follow a method similar to software engineering. That is building high-level designs on top of many levels of predictable, useful capabilities encapsulated in simple interfaces that allow the low-level capabilities to be improved.

The lowest level of structure is the individual microblocks that the product is assembled out of. A microblock will be a small fraction of the size of a human cell in molecular manufacturing. This is a convenient size to contain a basic functional unit such as a motor or CPU. A library of such microblock designs will be available for combining into larger functional units in future to cater to the needs of the molecular manufacturing field.

At the highest level, designers who are merely trying to recreate present level of product performance will find it easy to fit the required functionality into the product volume.

RESEARCH AND DEVELOPMENT

Prototype manufacturing of the products of molecular manufacturing is the first step. Manufacture of prototypes is a costly process. Rapid prototyping machines are essential for quicker development of the field. Presently, it is possible to make passive components and not integrated products.

High volume manufacturing may require overheads including expensive moulds, training of workers and costly raw materials. The design of products intended for high-volume manufacturing needs to take these factors into consideration. The product must be useful and also manufacturable. Designing the manufacturing process is an important part of the total design cost.

A nanofactory would be equally well-suited to rapid prototyping and high-volume production. Rapid testing of new product ideas and designs, more like compiling software as of now is possible than like present prototyping process.

Once a design is made and approved, production is the next step. Production site could be at point of sale (shops) or even points of use (home and offices) since the factories are smaller enough to carry anywhere. Warehousing and shipping may not be required, substantially reducing the expense and delay of product deployment. Efficient test marketing thereby reducing initial marketing costs is a possibility. Lower costs for research and development and far lower costs for initial deployment, would allow more innovation.

Nanoscale and microscale designs could be developed by a rapid genetic algorithm-type process with physical evaluation of the fitness function. An array of varied designs could be built and tested and the results used to refine the design parameters and also specify the next test planning. Rapid construction and testing of millions of components would allow physical implementation of genetic algorithm design methods. The products would not build copies of themselves and there would be no self-replication. The nanofactory would build each successive generation of products.

Cost of Manufacture of Nanofactory

A nanofactory capable of building a duplicate in an hour would be able to support a rapid increase in the number of nanofactories to any desired level. There would be no scarcity of manufacturing capacity.

Early nanofactories might require expensive feedstock and consume large amounts of power. But a combination of deliberate design processes and genetic algorithm approaches could produce rapid improvements in nanofactory component design. That will allow the use of simpler feedstock. Similarly nanofactory-built chemical processing plants can be developed, reducing the cost of feedstock. Because small designs could be built in large arrays, the processing system could use any convenient combination of mechanosynthesis, micro fluidics, and industrial chemistry.

Light-weight solar collectors could produce a plentiful supply of energy. If the products include solar collectors, then energy would not be limited. Thermionic solar cells have been built out of CVD diamond. Feedstock would be small carbon-rich molecules. Carbon is readily available anywhere on earth and nanofactory-built equipment might be used to process cheap carbon sources into feedstock. Nanofactories would not require special skills to operate and would not require working conditions that would be expensive to maintain.

With near-zero labour costs, low environment and capital costs and moderate energy and feedstock costs, the per-item cost of production will be very less regardless of device. This is possible once the initial bottlenecks in design, protypying and testing of marketing become successful.

Applications

A general-purpose manufacturing system capable of making complete high-performance products will have many applications. These include:

1.   massively parallel sensors and computers

2.   military force multiplication

3.   wholesale infrastructure replacement

4.   ecological footprint reduction and aerospace

Sensors   The ability to build kilograms of fully engineered nanoscale products if achieved, means that vast number of sensors could be produced at near-zero cost. These could be integrated into one structure for optical or medical data-gathering or could be incorporated in small distributed sensor platforms. More compact functionality, easier fabrication and more efficient use of power would give nano-built sensor platforms a significant advantage over MEMS technologies.

The ability to gather large amounts of physiological data (e.g. chemical concentrations or electrical potentials) in real time using a sensor array small enough to be inserted into the body without damage would be a huge help to medical research. Early and accurate detection of health conditions would help in the mitigation of many diseases. Early detection of adverse reactions to treatments would allow doctors to design more aggressive and experimental treatments without compromising safety. At one stage, it might be possible to bypass clinical trials entirely. At the other extreme, better understanding of causes, effects and feedback loops in the body would allow more subtle and less invasive treatment. Cell-sized machinery raises the possibility of cell-by-cell interventions, even genetic interventions, with more control and flexibility than current therapies. Interfacing to neurons, for both sensing and controlling neural signals, could be done more delicately and on a far larger scale than with today's electrode technology.

Computers   Massively parallel computers are expected to be one of the first products of a nanofactory development program. Applications of these computers include:

1. simulation

2. sensor array data processing

3. data-mining

4. pattern recognition

5. symbol manipulation

6. neural networks

7. precise construction could also be expected to be useful in building quantum computers.

Military applications   Portable high-volume general-purpose manufacturing of advanced products has got the possibility of greatly increasing the flexibility and power of military activities. Sensor and computer improvements would greatly improve situational awareness. Manufacture of products at the time and place where they are required would improve logistics and transportation of military equipment. The ability to build high-performance computers and actuators into any or every cubic millimetre of products would allow new kinds of weapons to be developed. The ability to inexpensively and rapidly build, test, and deploy new weapons would accelerate progress in weapon technology.

Wholesale infrastructure replacement   Low-cost, high-throughput manufacturing could be used to build very large quantities of products. Inefficient products, and even networks of inefficient products, could be replaced with little manufacturing effort. Of course replacement depends on many other factors including design effort, installation effort, political resistance, compatibility and consumer acceptance. Already available infrastructure deficiencies may provide many suitable targets for widespread deployment of nanofactory-built products.

Ecological footprint reduction   As the efficiency of infrastructure is improved and the ability to monitor the environment increases, it will be increasingly possible to reduce humanity's ecological footprint. Accidental or unnoticed damage could be reduced and the consequences of deliberate activities could be better known. Mineral extraction and processing activities, including fossil fuel consumption, could be reduced. Water could be used and re-used more efficiently. Widespread use of inexpensive greenhouses could save substantial amounts of water, topsoil, pesticides, fertilizer, and land area.

Aerospace   Aerospace hardware depends on high performance and light-weight materials. This is especially true for spacecraft. Orders of magnitude improvement in material strength, power density of actuators, and computer performance, make it reasonable to think of designing rockets with hardware mass of a small fraction of payload mass. Light-weight inexpensive hardware makes it easier to design combination airplane/rocket systems. It might even be worthwhile to include an efficient gas-processing system and fill a collapsible liquid oxygen tank after launch. Combination airplane/rocket systems capable of reaching orbit could be much smaller than rocket-only systems. Today, orbital access is expensive due to minimum rocket size, high construction cost, and the additional work required to avoid expensive failures. If spacecraft were smaller and failures were substantially less expensive, then research and development could proceed with less deliberation and more testing of advanced chemical and non-chemical designs.

MECHANOSYNTHESIS

Mechanosynthesis can be defined as any chemical synthesis that takes place by mechanical forces alone. In conventional chemistry, reactive molecules meet one another through random thermal motion in a liquid or vapour. In a hypothesized process of mechanosynthesis, reactive molecules would be attached to molecular mechanical systems and their encounters would result from mechanical motions bringing them together in planned sequences, positions and orientations. It is suggested that mechanosynthesis would avoid unwanted reactions by keeping potential reactants apart and would strongly favour desired reactions by holding reactants together in optimal orientations for many molecular vibration cycles. Mechanosynthetic systems would be designed in such a way to resemble some biological mechanisms.

History

The technique of moving single atoms mechanically was proposed by Eric Drexler in his book *The Engines of Creation* in 1986.

In 1988, researchers at IBM's Zürich Research Institute successfully spelled the letters "IBM" in Xenon atoms on a cryogenic copper surface, grossly validating the approach. Similar techniques to store computer data in a compact mode are produced. More recently the technique has been used to explore novel physical chemistries, sometimes using lasers to excite the tips to particular energy states or examine the quantum chemistry of particular chemical bonds.

In 2003, Oyabu *et al.* reported the first instance of purely mechanical-based covalent bond-making and bond-breaking, i.e., the first experimental demonstration of true mechanosynthesis with silicon atoms, not with carbon.

Methodology

Mechanosynthesis is building precise structures by the use of mechanical systems to control individual molecular reactions with atomic precision. The definition includes different kinds of fabrication operations and also types of control. It can occur in solution with part of the molecules or all molecules controlled. That is by using "machine phase" or "vacuum phase" chemistry. Mechanosynthesis can be utilized to add

1. small molecular fragments to selected positions on a large molecule,
2. selected monomers to the end of a polymer and
3. large molecular building blocks to a larger structure.

It can also be used for an opposite action namely to pull molecules apart or transfer atoms between molecules.

For reactions to occur the reactants are to be put closer and the energy barrier is to be overcome. Thermal motion can supply the energy needed. Piezochemistry which involves, applying hydrostatic pressure to affect reaction rates can be used. Light can also supply energy. Plasmons[4] might be useful to deliver optical energies more precisely.

Mechanosynthesis can reduce the rate of unwanted side reactions and allows a particular deposition site to be selected from among many chemically similar sites. Engineered heterogeneous products can be built by mechanosynthesis which is impossible by self-assembly or common solution chemistry.

In machine-phase chemistry, the lack of chemically active solvent simplifies computational chemistry simulations. Natural enzymes and antibodies can work without water or any solvent. A diverse set of machine-phase reactions are possible using scanning probe microscope.

Lack of solvent allows higher-energy molecules like the radicals to be used, that would not be stable in solution. The ability to create radicals in one process and then bring them together in a chosen orientation in a second and separate process can open new territories of chemistry. Creation of highly cross-linked materials like covalent solids are possible. The use of higher energy molecules can increase reliability of reactions.

Mechanochemistry

Mechanochemistry is the process in which molecules are fixed in single or more than one numbers. The molecules can be twisted or pulled or pushed in a precise way. This is to ensure a chemical reaction to occur in an appropriate and expected line. This reaction occurs in mechanochemistry in vacuum. No water molecules are involved in this reaction. The reaction is also fully controllable.

A mechanical manipulator is used for this purpose. The end of the mechanical manipulator is called "tool tip". To add an atom to a surface, the atom bound to a molecule namely "tool tip" is the starting point. The atom is then moved to the point where it should reach. The atom is moved next to the surface which is the reaching point of reaction.

The weaker bond of the atom is attached to the tool tip for easy release of the atom from the tool tip. When the tool tip and the surface of deposition of atom are brought nearer, the bond will get transferred to the surface. This is what is called normally acceptable reaction of chemistry. An atom always moves from one molecule to another molecule when the molecules come closer and the movement is energetically triggered.

In mechanochemistry, the tool tip molecule can be positioned by direct computer control and this reaction can be worked out at various sites on the surface.

Types of Reaction in Mechanosynthesis

The different types of reaction in mechanosynthesis are:

1. Formation of standard covalent bonds is one of the processes. This can be triggered by light, by electric fields or currents, by mechanical pressure or simply by holding reactive molecules near each other until thermal energy overcomes the reaction barrier. Weaker bonds, including hydrogen bonds and sulphur bonds, can link molecules large enough to include several bond sites.

2. In solvent, free-floating ions can play a part in the bonding. As an example, zinc coorinates tetrahedrally with cysteine and/or histidine amino acids, forming a fairly strong bond (Figure 6.6).

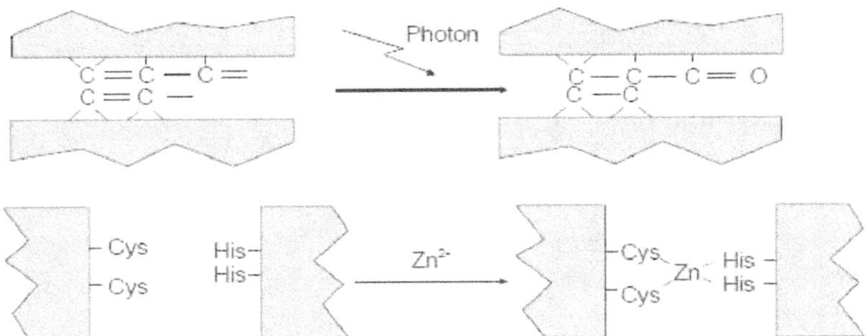

Figure 6.6 Zinc binding and photochemical binding are proposed for use in primitive molecular manufacturing systems

3. In vacuum or moderate-pressure gas, surface forces namely dispersion forces and van der Waals forces can hold things together quite tightly, up to 5% of the strength of diamond. Binding by surface forces is not actually a chemical reaction but surface forces form the weak end of a continuum. This molecular manufacturing design approach can apply to systems based on non-chemical fastening provided the parts being fastened are atomically precise and precisely placed.

Simulations of mechanosynthetic reactions have more predictive value than simulations of ordinary solvated reactions. Mechanosynthetic processes can physically constrain molecules in a way that avoids many unwanted reactions. Applying modest pressure to reactants can significantly alter the energetics of the reaction and thus shift reaction rates and equilibriums in desired directions. These advantages hold for both solvated and machine-phase mechanosynthesis. The ability to use extremely reactive species in machine-phase mechanosynthesis allows reactions in which the change in energy is far larger than in solution-phase chemistry.

Practical Fields of Mechanosynthesis

Mechanosynthesis as stated above has not yet been achieved. But primitive mechanochemistry has been performed at cryogenic temperatures using scanning tunnelling microscopes. Broader exploitation of mechanosynthesis awaits more advanced technology for constructing molecular machine systems including a molecular assembler or precursors thereof.

Mechanochemistry's potential use in automated assembly of molecular-scale devices is an important expectation. Such techniques appear to have many applications in medicine, aviation, resource extraction, manufacturing and warfare. Most theoretical explorations of such machines have focused on using carbon. Carbon is selected because of the many strong bonds it can form, the many types of chemistry these bonds permit and utility of these bonds in medical and mechanical applications. Carbon forms diamond and if cheaply available, would be an excellent material for many machines. Mechanosynthesis will be fundamental to molecular manufacturing based on nanofactories capable of building macroscopic objects with atomic precision. Research specifically aimed at positionally controlled diamond mechanosynthesis and diamondoid nanofactory development are on.

Getting one molecule to a known place on the microscope's tip is possible, but difficult to automate. Practical products need at least several hundred million atoms and this technique has not yet proven practical in forming a real product.

The Goals of Mechanoassembly Research

The following goals are aimed at:

1. Overcoming problems by calibration and selection of appropriate synthetic reactions—an assembler or molecular assembler fabrication. Once assemblers exist, geometric growth could reduce the cost of assemblers rapidly. Control by an external computer should then permit large groups of assemblers to construct large, useful projects to atomic precision. One such project would combine molecular-level conveyor belts with permanently mounted assemblers to produce a factory.

2. Dangerous industrial accidents as well as unexpected disastrous events equivalent to Chernobyl and Bhopal tragedies are to be kept in mind to avoid such problems out of mechanochemistry. The more remote issues of ecophagy, grey goo and green goo are to be studied regarding larger social and ecological implications for efficient and danger-free molecular manufacturing.

Current technical proposals for nanofactories do not include self-replicating nanorobots, and recent ethical guidelines would prohibit development of unconstrained self-replication capabilities in nanomachines.

Diamond Mechanosynthesis

Synthesizing diamond by mechanically removing/adding hydrogen atoms and depositing carbon atoms is known as diamond mechanosynthesis or DMS.

The DCB6Ge tooltip motif was the first complete tooltip ever proposed for diamond mechanosynthesis and remains the only tooltip motif that has been successfully simulaed for its intended function on a full 200-atom diamond surface. Figure 6.7 shows the mechanosynthetic reactions.

Figure 6.7 Mechanosynthetic reactions

Molecular Self-assembly

Modern synthetic chemistry has achieved the level of preparing small molecules to structure of any size. These methods are presently used to produce a wide variety of useful chemicals such as pharmaceuticals or commercial polymers. Extending this kind of control into supramolecular assemblies consisting of many molecules arranged in a well defined manner is the next step and is called molecular self-assembly.

A molecular assembler as defined by K. Eric Drexler is a proposed device able to guide chemical reactions by positioning reactive molecules with atomic precision. He also introduced a related term, "molecular manufacturing," which he defined as the programmed "chemical synthesis of complex structures by mechanically positioning reactive molecules, not by manipulating individual atoms. Some biological molecules such as ribosome fit this definition, since while working within a cell's environment, it receives instructions from messenger RNA and then assembles specific sequences of amino acids to construct protein molecules. However, the term "molecular assembler" usually refers to theoretical man-made or synthetic devices. They are thought to be highly desirable since they have been theorized to manufacture products with absolute precision and thus without any pollution. However, others have warned that such a powerful technology might escape human control and begin to compete with natural forms of life on earth.

Concept

The concepts of molecular self-assembly and/or supramolecular chemistry is utilized to automatically arrange themselves into some useful conformation through a bottom-up approach. The concept of molecular recognition is especially important. Molecules can be designed so that a specific conformation or arrangement is favoured due to non-covalent intermolecular forces. Two or more components can be designed to be complementary and mutually attractive so that they make a more complex and useful structure. Such bottom-up approaches should be able to produce devices in parallel and much cheaper than top-down methods. Most useful structures require complex and thermodynamically unexpected arrangements of atoms. Self-assembly based on molecular recognition in biology are Watson–Crick base-pairing and enzyme–substrate interactions. The challenge for nanotechnology is whether these principles can be used to engineer novel constructs in addition to natural ones.

Two main approaches are used in nanotechnology. In the "bottom-up" approach, materials and devices are built from molecular components which assemble themselves chemically by principles of molecular recognition. In the "top-down" approach, nano-objects are constructed from larger entities without atomic-level control. The impetus for nanotechnology comes from a renewed interest in Interface and Colloid Science, coupled with a new generation of analytical tools such as the atomic force microscope (AFM), and the scanning tunnelling microscope (STM). Combined with refined processes such as electron beam lithography and molecular beam epitaxy, these instruments allow the deliberate manipulation of nanostructures and have led to the observation of novel phenomena.

Synthetic assemblers have never been constructed and the possibility needs to be answered. Nanotechnology is an active area of research which has already been applied to the productio of real products. There are currently no research efforts into the actual construction of "molecular assemblers".

ENDNOTES

1. A solenoid is a three-dimensional coil. In physics, the term solenoid refers to a loop of wire, often wrapped around a metallic core, which produces a magnetic field when an electric current is passed through it. Solenoids are important because they can create controlled magnetic fields and can be used as electromagnets.

2. In quantum physics, the Heisenberg uncertainty principle states that locating a particle in a small region of space makes the momentum of the particle uncertain; conversely, that measuring the momentum of a particle precisely makes the position uncertain.

3. An actuator is a mechanical device for moving or controlling a mechanism or system.

4. In physics, a plasmon is a quantum of a plasma oscillation. The plasmon is the quasiparticle resulting from the quantization of plasma oscillations just as photons and phonons are quantizations of light and sound waves, respectively. Thus, plasmons are collective oscillations of the free electron gas density, oftenat optical frequencies. They can also couple with a photon to create a third quasiparticle called a plasma polariton.

REVIEW QUESTIONS

1. Define molecular manufacturing and its concepts. Differentiate MM from nanomanufacturing.

2. Discuss the approaches to MM.

3. Discuss in detail the goals of molecular manufacturing.

4. Discuss the laws governing the functioning of molecular manufacturing.

5. What are Scaling laws?

6. Write short notes on:
   i.   Information delivery in MM
   ii.   Errors and error handling of MM
   iii. Types of reactions in mechanosynthesis

7. What is mechanosynthesis?

8. What are the primary (basic) molecular manufacturing systems?

9. Write short notes on:

    i.   Tattoo type of architecture

    ii.   Silkscreen type of architecture

    iii.   Nanoblocks

10. Give an account of the development of basic molecular manufacturing systems.

11. Write in detail about high-performance nano- and microsystems.

12. What is a nanofactory?

13. Write short notes on:

    i.   Ridge joints

    ii.   Personal nanofactory

    iii.   Products of nanofactory

14. What is an advanced nanofactory?

15. What are the raw materials required in nanofactory?

16. Explain the nanofactory architecture.

17. Discuss in detail the molecular manufacturing products.

18. Give an account of high-performance materials and properties.

19. What are the applications of high-performance products?

20 Write short notes on:

    i.   Molecular self-assembly

    ii.   Nanorobotics

    iii.   *Engines of creation*

    iv.   Kim Eric Drexler

REFERENCES

1. Kim Eric Drexler. (1992). *Nanosystems: Molecular Machinery, Manufacturing, and Computation.*

2. Kim Eric Drexler, Chris, Peterson and Gayle Pergamit. (1991). *Unbounding the Future: the Nanotechnology Revolution.*

3. Gandall, B.C. (ed.). (1996). *Nanotechnology: Molecular Speculations on Global Abundance.* MIT Press, Massachusetts, Cambridge.

4. Robert, A. Freitas Jr., Ralph, C. Merkle. (2004). *Kinematic Self-Replicating Machines.* Landes Bioscience, Georgetown, TX.

Chapter 7

Nanomedicine and Nanobiology

Nanomedicine in simple terms is the medical application of nanotechnology. It encompasses fields like nanoparticle drug delivery and projected applications of molecular nanotechnology (MNT) and nanovaccine. Current problems for nanomedicine are related to toxicity and environmental impact of nanoscale materials and understanding of these issues. The other areas of nanomedicine are in conceptual, evolutionary, research and practical stages.

It is hoped that advances in nanomedicine will deliver a valuable set of research tools and clinically helpful devices. Expected applications are mainly in the pharmaceutical industry like advanced drug delivery systems, new therapies and also in *in vivo* imaging. Neuro-electronic interfaces and cell repair machines are the possible ones that could revolutionize medicine and the medical field. At the moment, nanomedicine is making its presence as one of the biggest industries in the world. Research is on especially in drug delivery by developing nanoscale particles or molecules to improve bioavailability. Drug delivery is being evolved for maximizing bioavailability related to specific places in the body and over a period of time. *In vivo* imaging is an important area of nanomedicine where tools and devices are being created. In imaging modalities such as ultrasound and MRI using nanoparticle contrast agents, required and favourable distribution of contrast agents in the body and also improved contrast picture of the lesions can be achieved. The new therapies and surgeries which are on the anvil can help to treat illnesses and diseases such as cancer. Using nanorobots such as neuro-electronic interfaces and cell repair machines, treating diseases is going to be easy.

BASIC CONCEPTS AND APPLICATIONS

The innovations in nanomedicine which are being developed will change the face of clinical medicine in future. These innovations include:

1.   drug delivery

2.   cancer diagnosis and therapy

3.   surgery

4.   *in vivo* therapy

5.   neuro-electronic interfaces

6.   cell repair machines

Drug Delivery

The pharmacological and therapeutic properties of drugs can be improved using designed drug delivery systems having lipid- or polymer-based nanoparticles. Drug delivery system is judged out of their ability to alter the pharmacokinetics and biodistribution of the drug. The unusual properties of nanoparticles can be used to improve drug delivery. Wherever larger particles are cleared from the body, cells take up these nanoparticles due to their size. Complex drug delivery mechanisms are under research, for getting drugs through cell membranes and into the cell cytoplasm. Many diseases depend upon processes within the cell and can only be impeded by drugs that make their way into the cell. Drug delivery systems need to be more efficient, and triggered response is one way for drug molecules to be used more efficiently. Drugs are placed in the body and get activated while encountering a particular signal. This is called triggered response. A drug delivery system in which both hydrophilic and hydrophobic environments exist, will improve the solubility of a drug with poor solubility and this system will replace the older ones. By using a drug delivery system with regulated drug release, tissue damage due to the drug can be eliminated. If a drug is cleared too quickly from the body, this could force a patient to use high doses. But with drug delivery systems, clearance can be reduced by altering the pharmacokinetics of the drug and so the dosage used can be minimal. Poor biodistribution is a problem that can affect normal tissues through widespread distribution in usual drug usage. But the particulates from drug delivery systems can lower the volume of distribution and reduce the effect on non-target tissue. Potential nanodrugs have got the possibility of working very specifically and by well-understood mechanisms. One of the major applications of nanotechnology and nanoscience in future will be in leading the development of completely new drugs with more useful behaviour and fewer side effects.

Cancer Diagnosis and Therapy

Nanoparticles of cadmium selenide (quantum dots) glow when exposed to ultraviolet light. When injected, they seep into cancer tumours. The surgeon can see the glowing tumour and use it as a guide for more accurate tumour removal. Figure 7.1 shows how nanoparticles or other cancer drugs might be used to trea cancer.

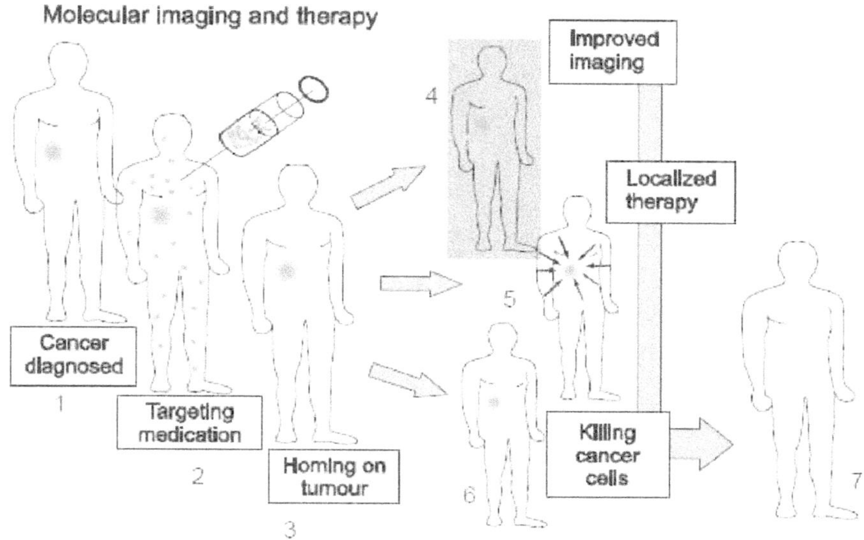

Figure 7.1   Action of nanoparticles in cancer treatments

When cancer is diagnosed in a person, the aim of therapy is targeting the medication towards the cancer cells only and is the ideal therapy of cancer. For that, the correct place or home of the tumour in the body is to be identified. By improved imaging due to nanotechnology and nanomedicine, the specific site or sites of cancer cells can be delineated or mapped. Using the nanomedical techniques, the cancer cells are targeted locally for treatment. So, the cancer cells are completely killed in the human body and the person is free from cancer.

Sensor test chips containing thousands of nanowires and able to detect proteins and other biomarkers left behind by cancer cells, could enable the detection and diagnosis of cancer in the early stages from a few drops of a patient's blood.

The nanoshells can be targeted to bond to cancerous cells by conjugating antibodies or peptides to the nanoshell surface. By irradiating the area of the tumour with an infrared laser, which passes through flesh without heating it, the gold is heated sufficiently to cause death to the cancer cells.

James Baker, scientist at University of Michigan, has discovered a highly efficient and successful way of delivering cancer-treatment drugs that are less harmful to the surrounding body. He has developed a nanotechnology that can locate and then eliminate cancerous cells. He used a molecule called dendrimer. This molecule has over a hundred hooks on it that allow it to attach to cells in the body for a variety of purposes. Folic acid, a vitamin which is received by cells in the body is easily attached to the few hooks in the dendrimers, to facilitate easy access to the cancer cells. Cancer cells have more vitamin receptors than normal cells, so Baker's vitamin-laden dendrimer will be absorbed by the cancer cell. To the rest of the hooks on the dendrimer, anti-cancer drugs are placed that will be absorbed along with the dendrimer into the cancer cell, thereby delivering the cancer drug to the cancer cell.

Surgery

A flesh-welder is used to fuse two pieces of chicken meat into a single piece. The two pieces of chicken are placed together so that they are closely aligned. A greenish liquid containing gold-coated nanoshells is dribbled along the seam. An infrared laser is traced along the seam, causing the two sides to weld together. This could solve the difficulties and blood leaks caused when the surgeon tries to restitch the arteries he/she has cut during a kidney or heart transplant. The flesh welder could mend the artery into a perfect seal. (Rice University research finding).

In vivo Therapy

Nanodevices   Nanodevices with their components manufactured using ^{13}C atoms rather than the natural ^{12}C isotope of carbon, could be observed at work inside the body using MRI, since ^{13}C has a nonzero nuclear magnetic moment. The methodology for nanodevices is as follows: Medical nanodevices would be injected into a human body that would go to work in a specific organ or tissue mass. The device will be monitored for the progress, and will be verified that the nanodevices are in the correct target treatment region. When a part of the body is scanned, the nanodevices congregated neatly around their target like a tumour mass. Thus the procedure's successfulness can be verified.

Tracking movement   Tracking movement helps to determine drug distribution or metabolism of substances. The cells are dyed so that they can be tracked throughout the body. These dyes are excited by light of a certain wavelength to emit light. While different colour dyes absorb different frequencies of light, there was a need for as many light sources as cells. For this, luminescent tags (Figure 7.2) are used. These tags are quantum dots attached to proteins that penetrate cell walls. The dots can be random in size, can be made of bio-inert material. This bio-inert material demonstrates the size-dependent nanoscale property in showing colour changes. The sizes of the material used are selected in such a way that the frequency of light used for making a group of quantum dots fluoresce can also be used to make another group of quantum dots lighted. Because of this methodology, a single light source can be used to lit both grups of quantum dots.

Figure 7.2 Luminescent tag

Photodynamic therapy   In photodynamic therapy, a particle is placed inside the body and is illuminated using outside source of light. The light gets absorbed by the particle and if the particle is metal, energy from the light will heat the particle and surrounding tissues. Light may also be used to produce high-energy oxygen molecules which will chemically react with and destroy most organic molecules that are next to them like tumours. This therapy is appealing for many reasons. It leaves no "toxic trail" of reactive molecules throughout the body as in routine chemotherapy, because it is directed where only the light is shone and the particles exist. Photodynamic therapy has potential for a noninvasive procedure for dealing with diseases, growths, and tumours.

Neuro-electronic Interfaces

Neuro-electronic interfaces are a system in which nanodevices are operating through computers which are joined and linked to the nervous system. A molecular structure that permits control and detection of nerve impulses by an external computer needs to be built up for this purpose. The computers have to interpret, register, and respond to signals given by the body. The demand for such structures is huge because many diseases involve the decay of the nervous system (amyotrophic lateral sclerosis and multiple sclerosis). Injuries and accidents impairing the nervous system resulting in dysfunctional systems and paraplegia also need such arrangements. If computers could control the nervous system through neuro-electronic interface, problems that impair the system could be controlled so that effects of diseases and injuries could be overcome. Regarding power source for such a system to function, two options are available. They are refuellable and nonrefuellable strategy. A refuellable strategy implies that energy is refilled continuously or periodically with external sonic, chemical, tethered, or electrical sources. A nonrefuellable strategy implies that all power is drawn from internal energy storage which would stop when all energy is drained. Figure 7.3 shows the neurn–semiconductor interface.

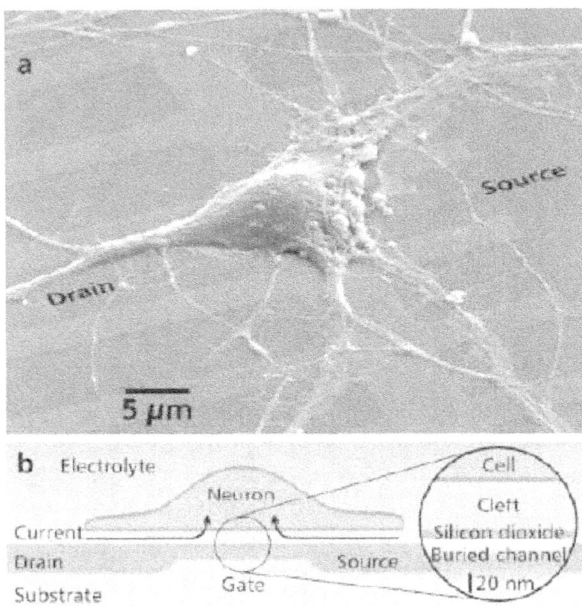

Figure 7.3   Neuron–semiconductor interface (*Source*: Neurophilosophy.wordpress.com)

Cell Repair Machines

Using drugs and surgery, tissue repair is enhanced but nature will take its own time to heal and the tissues will have to repair on their own. This goes by the dictum that "Surgeon helps nature to heal the wounds". Using molecular machines, more direct repairs are possible. Cell repair machines will utilize the same tasks that living systems already prove possible. Access to cells is possible through sticking needles into cells without killing them. By means of this methodology, molecular machines are capable of entering the cell. All specific biochemical interactions show that molecular systems can recognize other molecules by touch. They can build or rebuild every molecule in a cell as well as disassemble damaged molecules. Finally, cells that replicate give the conclusion that molecular systems can assemble every system found in a cell. Nature has demonstrated the basic operations needed to perform molecular-level cell repair. In future, using the same type of activities, nanomachine-based systems, which are able to enter cells, sense differences from healthy ones and make modifications to the structure of the cells, will be built.

Cell repair machines (Figure 7.4) will be more complex compared to the size of viruses or bacteria because of their compact parts. In the evolution of these machines, the early machines will be able to correct a single molecular disorder like DNA damage or enzyme deficiency. As they evolve, in later stages, cell repair machines with the help of advanced artificial intelligence systems, will be programmed for more abilities or repairs. These ae speculations at the moment.

Figure 7.4   Nanotech cell repair machines

Nanocomputers will be needed to guide these machines. These computers will direct machines to examine, take apart and rebuild damaged molecular structures. Repair machines will be able to repair whole cells by working structure by structure. Then by working cell by cell and tissue by tissue, whole organs can be repaired. Finally, by working organ by organ, health is restored to the body. Cells damaged to the point of inactivity can be repaired because of the ability of molecular machines to build cells from scratch. Due to the self-repair mechanism, it is expected that cell repair machines will heal diseases without the help of mdicines in fuure.

Cell Reair→Tissue Repair→Organ Repair→Whole Body Repair

A new wave of technology and medicine is being created and its impact on the world is going to be monumental and nature-beating. From the possible applications such as drug delivery and *in vivo* imaging to the potential machines of the future, advances in nanomedicine are being made every day in a stepwise pattern.

NANOBIOTECHNOLOGY

The twin fields nanotechnology and biotechnology have got certain common platforms of working in the arena of applications which has revolutionized the biological investigations. Structural and surface properties of nanostructures can be modified systematically. This results in modification of the properties of nanostructures at a nanoscale level. This makes them very useful in biological contexts, from fundamental scientific studies to commercially viable technologies. Nanostructures with highly controlled properties in the nanometre size range are available in a progressive manner now. This has created interest in their use in biotechnological systems.

"Size does matter" is of paramount importance in nano and biotechnology. Nanoparticles are similar in size range to many common biomolecules. This makes them natural companions in hybrid systems. The new and unique properties that nanostructures bring to biotechnological applications are very much important. By controlling structure precisely at nanoscale dimensions, it is possible to control and tailor properties of nanostructures, such as semiconductor nanocrystals and metal nanoshells, in a very accurate manner. It is possible to make modifications to nanostructures for better integration with biological systems. Modifying their surface layer for enhanced aqueous solubility, biocompatibility or biorecognition, help in their applications in biological systems. With selected biomolecules bound to nanostructure surfaces, new 'hybrid' nanostructures can be obtained for applications such as biosensing and imaging. Nanostructures can be embedded in other biocompatible materials to modify material properties or to impart new functionality.

The combination of biological and non-biological systems at the nanoscale level is not a new concept. The bioconjugate chemistry's broad field is based on combining the functionalities of biomolecules and non-biologically derived molecular species for specialized uses. Its applications range from markers for research in cell and molecular biology to biosensing, bioimaging and masking of immunogenic moieties to targeted drug delivery.

Most of the current applications of nanostructures in biotechnology are a natural evolution of this approach. Several of the 'breakthrough' applications which are recently demonstrated using nanostructure-biomolecular hybrids are traditional applications. They are initially addressed by standard molecular bioconjugate techniques that have been revitalized with these newly designed nanostructure hybrids. The reasons for replacing conventional molecular tags like fluorescent chromophores with nanostructures are well known. Nanostructures possess properties far superior to the molecular species they replace, like higher quantum efficiencies, greater scattering or absorbance cross sections, optical activity over more biocompatible wavelengths, and better chemical or photochemical stability. The systematic control of nanostructure properties obtained by controlled variations in particle size and dimension is an advantage over molecular tags, properties of which vary nonsystematically between molecular species. This systematic variation of properties via structure variation improves traditional applications and leads to new, unique applications than the conventional molecular bioconjugates. An example is the optical properties of semiconductor nanocrystals and metal nanoshells, which are new and robust fluorophores, absorbers and scatterers in the near infrared, a region of the electromagnetic spectrum where tissue is essentially transparent. These new nanostructures will greatly facilitate new *in situ* probes and sensor methods.

Several successful examples of nanostructures that have been integrated with biomolecular species and applied to relevant problems in biotechnology are described under the following categories.

1. The use of bioconjugate semiconductor nanocrystals, or "quantum dots", as fluorescent biological labels.

2. A new and powerful assay based on the optical properties of bioconjugate gold nanoparticles, a new innovation on a traditional bioconjugate technology.

3. The biotechnologically friendly properties of gold nanoshells.

4. A novel photo-thermally triggered drug delivery system based on a new nanoshell—polymer composite.

Fluorescent Biological Labels

Semiconductor nanocrystals are highly light-absorbing, luminescent nanoparticles. Their absorbance onset and emission maximum will shift to higher energy with decreasing particle size, due to quantum confinement effects. These nanocrystals are 2–8 nm in diameter. Molecular fluorophores have very narrow excitation spectra. But semiconductor quantum dots absorb light over a very broad spectral range. This makes it possible to optically excite a broad spectrum of quantum dot colours using a single excitation laser wavelength, which enables one to simultaneously probe several markers. The luminescence properties of semiconductor nanocrystals have been sensitive to their local environment and nanocrystal surface preparation. Recent core-shell geometries where the nanocrystal is encased in a shell of a wider band gap semiconductor have resulted in increased fluorescence quantum efficiencies and greatly improved photochemical stability.

The nanoparticles can be made biocompatible by using both the silica layer and the covalent attachment of proteins to the mercaptoacetic acid coating. Specific binding to cell surfaces, insertion into cells, and binding to cell nuclei have all been demonstrated following conjugation of the nanoparticle with the appropriate targeting protein.

Colorimetric Assay

The use of gold colloid in biological applications dates back to 1971, when Faulk and Taylor invented the immunogold staining procedure. From then on, the labelling of targeting molecules, mainly proteins, with gold nanoparticles has revolutionized the visualization of cellular or tissue components by electron microscopy. The optical and electron beam contrast qualities of gold colloid have provided excellent detection qualities for such techniques as immunoblotting, flow cytometry, and hybridization assays. Conjugation protocols exist for the labelling of a broad range of biomolecules with gold colloid, such as protein A, avidin, streptavidin, glucose oxidase, horseradish peroxidase, and IgG.

Gold nanoparticle conjugation was recently applied to polynucleotide detection. It involves the change in optical properties resulting from plasmon–plasmon interactions between locally adjacent gold nanoparticles. The characteristic red colour of gold colloid changes to a bluish purple colour upon colloid aggregation. In the case of polynucleotide detection, mercaptoalkyloligonucleotide-modified gold nanoparticle probes are used. When a single-stranded target oligonucleotide is introduced into solution, a polymer network will be formed consisting of the target oligonucleotide and the conjugated nanoparticles. This condensed network brings the nanoparticles into close vicinity that is enough to induce a dramatic red-to-blue macroscopic colour change. Because of the extremely strong optical absorption of gold colloid, this colorimetric method can be used to detect ~10 fmol[1] of an oligonucleotide, which is 50 times more sensitive than sandwich hybridization detection methods based on fluorescence detection.

Drug Delivery System

Gold nanoshells are new composite nanoparticles. They combine infrared optical activity with the uniquely biocompatible properties of gold colloid. Metal nanoshells are concentric (spherical) nanoparticles consisting of a dielectric (typically gold sulphide or silica) core and a metal (gold) shell. By varying the relative thickness of the core and shell layers, the plasmon-derived optical resonance of gold can be dramatically shifted in wavelength from the visible region into the infrared. Infrared wavelength range is the region of highest physiological transmissivity. By varying the absolute size of the gold nanoshell, it can be made to either selectively absorb (for particle diameters <~75 nm) or scatter incident light. Because the gold shell layer is deposited using the same chemical methods used to grow gold colloid, the surface properties of gold nanoshells are virtually identical to those of gold colloid. The same conjugation protocols used to bind a wide variety of biomolecules to gold colloid is transferable to gold nanoshell conjugation.

Successful gold nanoshell conjugation with enzymes and antibodies has been demonstrated. Gold nanoshells also take advantage of the inherent biocompatibility of gold, not requiring further surface functionalization or protective layer growth. When optically absorbing gold nanoshells are embedded in a matrix, illuminating them at their resonance wavelength causes the nanoshells to transfer heat to their local environment.

This photothermal effect can be used to optically "remote control" drug release in a nanoshell-polymer composite drug delivery material. Copolymers of *N*-isopropylacrylamide (NIPAAm) and acrylamide (AAm) exhibit a critical solution temperature (LCST) that is slightly above body temperature. When the temperature of the copolymer exceeds the LCST, the material collapses causing a burst release of any soluble material held within the polymeric matrix.

NANOBIOTECHNOLOGICAL DEVICES

Many nanodevices are useful in the study and application of nanobiotechnology.

Nanoparticles

Nanoparticles such as **quantum dot** nanocrystals are of the size of a protein molecule or short stretch of DNA. Quantum dots can be engineered to absorb and emit many wavelengths of light with very sharp precision. This makes them ideal for protein–protein interaction studies as they can be linked to molecules to form long-lived probes. They can track biological events by tagging specific proteins or DNA in order to follow their progress through biological pathways. In medicine, quantum dots could be used for diagnostic purposes.

Dendrimers

Dendrimers are synthetic molecules with nanostructure and are a few nanometres wide. They have a central core with regular branching structures. Each synthetic step of formation of dendrimer produces one layer and the number of layers and size of the dendrimer depend on the number of synthetic steps. The outer layer of dendrimer is engineered to have specific functional groups acting as hooks for specific binding of molecules such as DNA. In gene therapy, for delivering DNA into the cell, dendrimers are useful. Viral vectors when used in gene therapy trigger immune response. Dendrimers are not expected to have such response in principle. In other words, it should be immunocompatible.

Nanorobots

Research carried out in the field of mechano-compatibility for the human body in the past few years is to create a nanorobot within the human body which can operate both chemically and physically. Biomedical instruments and procedures used by physicians to explore tissues and cells are available but a new field of interest in exploring the human body with nanorobots is on the anvil. These nanorobots must pass through a person's body without causing excessive bruising, itching, and other disturbances. The key for success in this field of study is finding a way nanorobots can perform the maximum amount of biomedical operations with the least amount of irritation and other possible illnesses to the patient.

Nubot

Nubot is an abbreviation for "Nucleic Acid Robots." Nubots are synthetic robotic devices at the nanoscale. Representative nubots include the several DNA walkers.

Potential applications

- Early diagnosis and targeted drug delivery for cancer
- Biomedical instrumentation
- Surgery
- Pharmacokinetics

Nanoshell

A spherical core of a particular compound surrounded by a shell a few nanometres in thickness is called as nanoshell (core-shell).

Nanoshell applications are very much worked out in the biological field. By altering the thickness of the shells, nanoshells with high absorptions at biologically useful wavelengths are created. Infrared region of wavelength is the useful spectrum because, the tissues absorb less in infrared region. In other words, tissue transmissivity increases in this region.

Metal nanoshells are a type of composite spherical nanoparticle with a dielectric core like silica covered by an ultrathin metallic shell which is typically gold. They possess optical and chemical properties for biomedical imaging and therapeutic applications in a suitable and favourable way. The optical resonance of these nanoparticles can be precisely and systematically varied over a broad region ranging from the near-UV to the mid-infrared by varying the relative dimensions of the core and the shell. This range includes the near-infrared (NIR) wavelength region where tissue transmissivity peaks. In addition to spectral tunability, nanoshells offer other advantages over conventional organic dyes including improved optical properties and reduced susceptibility to chemical/thermal denaturation. Further, the same conjugation protocols used to bind biomolecules to gold colloid are easily modified for nanoshells.

Recently, interest has developed in the creation of nanotechnology-based platform technologies which couple early molecular-specific detection strategies with appropriate therapeutic intervention and monitoring capabilities.

Gold nanoshell with a dielectric core (gold sulphide, silicon, etc.) (Figure 7.5) is the much sought-after metal in this application. Gold is a biocompatible compound, making it a useful material for medical applications. Gold (Au) nanoshells exhibit tunable plasmonic resonance based on the ratio of the shell to core thickness. In addition, the overall size of such aparticle allows independent control of its scattering cross section.

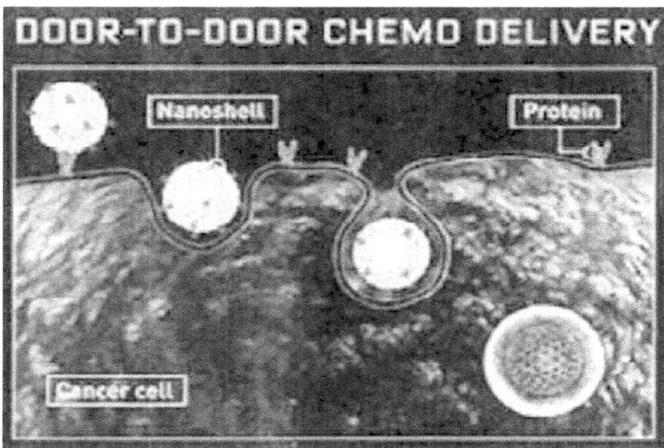

Figure 7.5 Nanoshells—precision-guided tumour gold nanoshell with silica core

The overall size of silica/gold nanoshells may make clearance of these particles through renal system. For removal from the body via the kidneys (renal system), a particle's hydrodynamic diameter must be approximately less than 6 nm. Long-term *in vivo* toxicity studies must be undertaken to determine the toxicity of any nanoparticle.

Nanoshells are currently being tried as:

1. A treatment for cancer similar to chemotherapy but without the toxic side-effects

2. Inexpensive, quick analysis of "samples as small as a single molecule" and depends on Raman spectroscopy

Gold nanoshells possess physical properties similar to gold colloid, in particular, a strong optical absorption due to the collective electronic response of the metal to light. The optical absorption of gold colloid yields a brilliant red colour which has been of considerable utility in consumer-related medical products, such as home pregnancy tests. In contrast, the optical response of gold nanoshells depends dramatically on the relative size of the nanoparticle core and the thickness of the gold shell. By varying the relative core and shell thicknesses, the colour of gold nanoshells can be varied across a broad range of the optical spectrum that spans the visible and the near infrared spectral regions (Figure 7.6). Gold nanoshells can be made to either preferentially absorb or scatter light by varying the size of the particle relative to the wavelength of the light at their optical resonance. It is now possible to predictively design gold nanoshells with the desired optical resonant properties and fabricate the nanoshell with the dimesions and nanoscale tolerances necessary to achieve these properties.

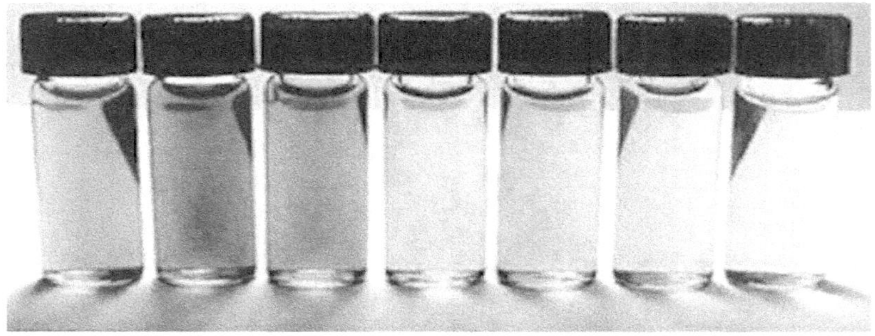

Figure 7.6 Visual demonstration of the tunability of metal nanoshells

Combining advances in biophotonics and nanotechnology offers the opportunity to significantly impact future strategies towards the detection and therapy of cancer. Today, cancer is typically diagnosed many years after it has developed usually after the discovery of either a palpable mass or based on relatively low resolution imaging of smaller but still significant masses. In the future, it is likely that contrast agents targeted to molecular markers of disease will routinely provide molecular information that enables characterization of disease susceptibility long before pathologic changes occur at the anatomical level. Currently, our ability to develop molecular contrast agents is at times constrained by limitations in our understanding of the earliest molecular signatures of specific cancers. Although the process of identifying appropriate targets for detection and therapy is ongoing, there is a strong need to develop the technologies which will allow us to image these molecular targets *in vivo* as they are elucidated. Combining the tools of both fields—together with the latest advances in understanding the molecular origins of cancer—may provide a fundamentally new approach to detection and treatment of cancer. The nanoshells can be targeted to bond to cancerous cells by conjugating antibodies or peptides to the nanoshell surface. By irradiating the area of the tumour with an infrared laser, which passes through flesh without heating it, the gold is heated sufficiently to cause death to the cancer cells.

Researchers at Rice University under Prof. Jennifer West have demonstrated the use of 120 nm diameter nanoshells coated with gold to kill cancer tumours in mice.

OTHER APPLICATIONS OF NANOBIOTECHNOLOGY

1. Targeted nanoparticles incorporating siRNA is considered for cancer treatment.

2. Characterization and *in vivo* efficacy of targeted nanoparticles for systemic siRNA delivery to tumours.

3. Nano-neuro knitting using nanotechnology is used to repair the brain.

4. Automated 3D cortex image data acquisition is used for identification of synaptic regions with their characteristic neurotransmitter-carrying vesicles.

5. Developing super-paramagnetic nanoparticles for central nervous system axon regeneration.

6. Nanowire array technologies for investigation of neural activity.

7. Bioactive aligned nanofibres for nerve regeneration.

8. Nanomedicine as a way for nerve cell regeneration.

9. Nanotechnology treatment for pancreatic cancer.

BIOSENSORS

A device that detects, records, and transmits information regarding a physiological change or process is called as biosensor. A device that uses biological materials to monitor the presence of various chemicals in a substance can also be called as biosensor. Biosensors play a vital and important role in medicine, industry and the environment, providing routine analysis, crucial monitoring, and early detection of problems and crisis points. Biosensors are useful in the diagnosis and monitoring of diseases, drug discovery, proteomics and the environmental detection of pollutants and biological agents. The development of biosensors for the above purposes is an extremely significant problem in biomedical research. Basically, a biosensor is derived from the coupling of a ligand–receptor binding reaction to a signal transducer. Various signal transduction methods include optical, radioactive, electrochemical, piezoelectric, magnetic, micromechanical, and mass spectrometric methods. Biosensor research has been mainly devoted to the evaluation of the relative merits of these signal transduction methods. The development of large-scale biosensor arrays composed of highly miniaturized signal transducer elements is an ongoing research. These elements enable the real-time, parallel monitoring of multiple species and are an important goal in biosensor research.

Development of optical biosensors and chemosensors based on the extraordinary optical properties of noble metal nanoparticles is an alternative strategy in biosensor research.

The history of biosensors started in 1956, when Leland C Clark Jr., known as the father of biosensors, published his definitive paper on the oxygen electrode.

Definition

A biosensor is a device for the detection of an analyte that combines a biological component with a physicochemical detector component. It combines a biochemical recognition/binding element (ligand) with a signal conversion unit (transducer).

A biosensor consists of three parts:

1. The *sensitive biological element* which may be i) biological material, e.g. tissue, microorganisms, organelles, cell receptors, enzymes, antibodies, nucleic acids, synthetic receptors, sensing organs, etc. ii) a biologically derived material or iii) biomimic. The sensitive elements can be created by biological engineering.

2. The *transducer* in between (associates both components). The transducer acts as an interface, measuring the physical change that occurs with the reaction at the bioreceptor and then transforming that energy produced by the biochemical interactions into measurable electrical output. Physical transducers include optical, electrochemical, opto-electronic, piezoelectric, magnetic, thermal, and mass transducers.

3. The *detector element* which works in a physico-chemical way namely optical, piezoelectric electrochemical, thermometric, or magnetic. Signals from the transducer are passed to a microprocessor where they are amplified and analysed. The data is then converted to concentration units and transferred to a display or/and data storage deice.

A schematic flowchart showing the main components of a biosensor is shown in Figure 7.7.

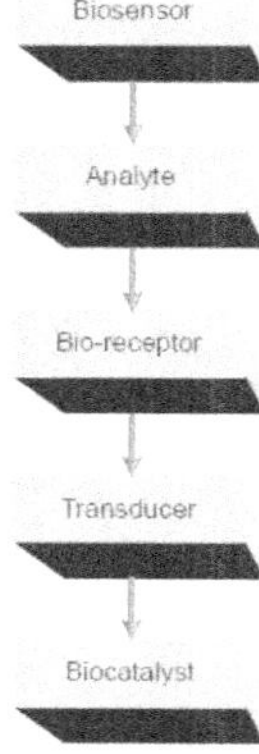

Figure 7.7 Components of a biosensor

The output from the transducer is amplified, processed and displayed.

Blood glucose biosensor (Figure 7.8) is the most widespread example of a commercial biosensor. This uses an enzyme to break blood glucose down. In doing so, it transfers an electron to an electrode and this is converted into a measure of blood glucose concentration. Because of the epidemic nature of diabetes mellitus as a whole in the world and specifically in India, the demand for such sensors is increasing. That has fuelled the development of associated sensor technologies. Of late cholesterol and triglycerides are also estimated in the same way as glucose. Even the HbA1c which is a measure of the control level of diabetes for the previous three months of a patient's life is possible to be assessed by the biosensor technology in a simple way. This helps the doctor to advice the patient during the initial visit itself without waiting for the results to arrive from the laboatory. Like these, so many instant result kits for medical investigations are in the pipeline.

Figure 7.8 Blood glucose biosensor

The pattern of response from the detectors in electronic nose devices consisting of arrays of many different detector molecules is used to fingerprint a substance.

A canary in a cage, as used by miners to warn of gas could be considered as a biosensor. Many of today's biosensor applications are similar in nature. They use organisms which respond to toxic substances at a much lower level than detected ordinarily to warn of their presence. Devices of such nature are useful both in environmental monitoring and in water treatment facilities.

Principles of Detection

The transducer converts the biochemical interactions into a measurable electronic signal (Table 7.1). Electrochemical, electro-optical, acoustic and mechanical transducers are among the many types found in biosensors. The transducer works either directly or indirectly.

Biosensors are also characterized by their specificity, or their ability to recognize a single compound among other substances in the same sample. The selectivity of biosensors is determined by both he bioreceptor and the method of transduction.

Table 7.1 Biosensors and their measurable signals

Biosensors	Measures changes in
Piezoelectric	Mass
Electrochemical	Electric distribution
Optical	Light intensity
Calorimetric	Heat

Biosensors are of two types.

1. *Direct detection biosensors*   Direct recognition sensors are the ones in which the biological interaction is directly measured using non-catalytic ligands such as cell receptors or antibodies.

2. *Indirect detection biosensors*   This class of transducers called indirect detection sensors mainly relies on secondary labelled elements namely fluorescently tagged antibodies or catalytic elements such as enzymes.

Optical (Photometric) Biosensor

A sensor that uses light to detect the effect of a chemical on a biological system is an optical biosensor. Optical biosensors based on the phenomenon of surface plasmon resonance are evanescent wave techniques. This utilizes a property shown by gold and other materials; specifically that a thin layer of gold on a high refractive index glass surface can absorb laser light, producing electron waves (surface plasmons) on the gold surface. This occurs only at a specific angle and wavelength of incident light and is highly dependent on the surface of the gold, such that binding of a target analyte to a receptor on the gold surface produces a measurable signal.

Surface plasmon resonance sensors operate using a sensor chip consisting of a plastic cassette supporting a glass plate, one side of which is coated with a microscopic layer of gold. This side contacts the optical detection apparatus of the instrument. The opposite side is then contacted with a microfluidic flow system. The contact with the flow system creates channels across which reagents can be passed in solution. This side of the glass sensor chip can be modified in a number of ways, to allow easy attachment of molecules of interest. Normally it is coated with carboxymethyl dextran or similar compound.

Light at a fixed wavelength when passed off the gold side of the chip, it is reflected at the angle of total internal reflection and the light so reflected is detected inside the instrument. The light so detected in turn induces the evanescent wave to penetrate through the glass plate. This in turn passes into the liquid flowing over the surface. The refractive index of the light at the flow side of the chip surface depends on the light reflected off the gold side. Materials bound to the flow side of the chip alters the refractive index. So, by calculating or measuring the refractive index changes, biological interactions can be measured to a high degree of sensitivity.

Other optical biosensors are mainly based on changes in absorbance or fluorescence of an appropriate indicator compound.

Electrochemical Biosensor

Electrochemical biosensors are normally based on enzymatic catalysis of a reaction that produces ions. The sensor substrate contains three electrodes—a reference electrode, an active electrode and a sink electrode. A counter electrode may also be present as an ion source. The target analyte is involved in the reaction that takes place on the active electrode surface, and the ions produced create a potential which is subtracted from that of the reference electrode to give a signal.

The potentiometric biosensors, are similar to electrical biosensors They are screen-printed, conducting-polymer coated, open-circuit potential biosensors based on conjugated polymers immunoassays. They have only two electrodes and are extremely sensitive, robust and accurate. They enable the detection of analytes at levels, previously achievable only by HPLC and LC/MS, and without rigorous sample preparation. The signal is produced by electrochemical and physical changes in the conducting polymer layer due to changes occurring at the surface of the sensor. Such changes can be attributed to ionic strength, pH, hydration and redox reactions, and the latter due to the enzyme label turning over a substrate.

Others

Piezoelectric sensors utilize crystals which undergo an elastic deformation when an electrical potential is applied to them. An alternating potential (AC) produces a standing wave in the crystal at a characteristic frequency. This frequency is highly dependent on the surface properties of the crystal, such that if a crystal is coated with a biological recognition element, the binding of a large target analyte to a receptor will produce a change in the resonance frequency, which gives a binding signal. This is a special application of the quartz crystal microbalance in biosensor. Thermometric and magnetic based biosensors are rare.

Applications

There are many potential applications of biosensors of various types. The main requirements for a biosensor approach, to be valuable in terms of research and commercial applications are the identification of a target molecule, availability of a suitable biological recognition element, and situations in which instead of sensitive laboratory-based techniques, disposable and portable detection systems are preferred, have to be identified.

Some examples are given below:

◈ Glucose monitoring in diabetes patients—historical market driver

◈ Other medical health-related targets

◈ Environmental applications, e.g. the detection of pesticides and river water contaminants

◈ Remote sensing of airborne bacteria, e.g. in counter-bioterrorist activities

◈ Detection of pathogens

◈ Determining levels of toxic substances before and after bioremediation

◈ Detection and determining of organophosphate

◈ Routine analytical measurement of folic acid, biotin, vitamin B_{12} and pantothenic acid as an alternative to microbiological assay

◈ Determination of drug residues in food, such as antibiotics and growth promoters, particularly meat and honey

◈ Drug discovery and evaluation of biological activity of new compounds

◈ Cancer diagnostics

NANOBIOSENSORS

A device that detects, records, and transmits information regarding a physiological change or process is called as biosensor. Biosensors which use nanotechnology are called nanobiosensors. Nanobiosensors certainly have their place in the futuristic universe of the nanotechnologies. One of the important areas of nanobiosensors is the development of nanonose. Nanobiosensors are capable of detecting odours at levels which may be imperceptible to the human nose. Currently there are several hypothesized ways to produce nanosensors. They are

1.   Top-down method
2.   Bottom-up method
3.   Self-assembly

Nanonose

Using smell to detect the early warning signs of different illnesses are possible with the technology that replicates and improves upon the human olfactory system namely tiny bioelectronic sensors. Modern-day doctors can soon start using such devices for detection of diseases.

This new interdisciplinary technology approach will ultimately lead to electronic noses based on natural olfactory receptors that could be used in health care, agriculture, industry, environmental protection or security. Nanobiosensors to mimic the way human and animal noses respond to different odours are developed.

The latest generation of electronic noses are nanobiosensor arrays based on the electrical properties of olfactory receptors from mammals, which are proteins, situated in the olfactory epithelium within the nasal cavities.

By placing a layer of proteins that constitute the olfactory receptors in animal noses on a microelectrode and measuring the reaction when the proteins come into contact with different odorants, the system is capable of detecting odorants at concentrations that would be imperceptible to humans. Figure 7.9 shows a nanonose.

Several hundred different proteins would be needed for an electronic nose to detect any smell because different proteins react to different odorants and it is the resultant combination of reactions that identifies a certain smell. Nanotechnology makes such an electronic nose feasible evn though the human nose uses 1,000 different proteins to allow the brain to recognize 10,000 different smells.

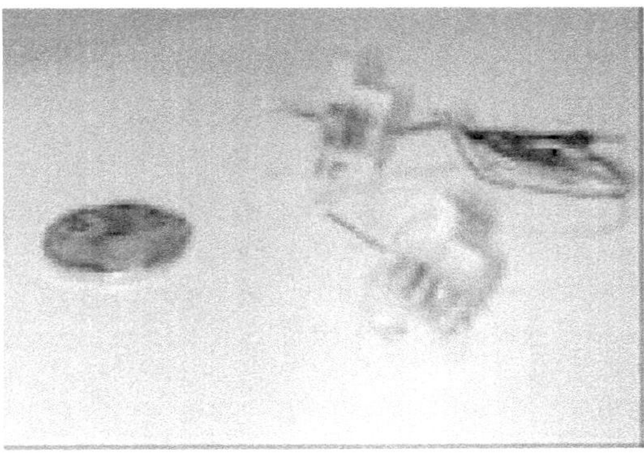

Figure 7.9   Nanonose

Nanobiosensors can react to a few molecules of odorant with a very high degree of accuracy. These tiny bioelectronic sensors, represent a revolution in smell technology and an example of a biomimetic devices obtained by converging nano-bio-info technologies.

To design this artificial biosensor system, researchers used the physical reactions which occur in animals when an odorant molecule is detected by an olfactory receptor. When an odorant molecule binds to the olfactory receptor, this causes a conformational change and modifications to the impedance-metric properties of the protein, which are recorded via nanoelectrodes.

Modifications in the receptors would result from the specific behaviour of the olfactory receptor in response to odorant stimulation. Nanobiosensors can react to a very small amount of odorant molecules (high level of sensitivity) with a considerable degree of precision (high specificity). Nanobiosensors were produced by grafting these specific proteins, carried by nanovesicles, between two nanoelectrodes.

To develop a bioelectronic nose capable of detecting and identifying a wide variety of odours, researchers can use yeast cells capable of producing a broad range of different olfactory receptors, or from the rat or dog (both species are endowed with more than a thousand different receptors). Combining the responses of these receptors make it possible to detect any odour. Research and development of the instrumentation and software tools necessary for an electronic nose to recognize smells and the role played by the brain in the olfactory system are under creation.

Electronic instrumentation capable of performing electrical measurements at the nanoscale level has been developed and adapted to an atomic force microscope with atofarad (10^{-15}) precision. This can help to create devices for diagnosing organ failure, bacterial infections or diseases such as cancer. A major challenge is the establishment of a precise odorant disease signature, understood as the mix of volatile compounds whose concentration in a body fluid (i.e., urine, blood, pus, etc.) or in the breath varies in patients with the malignancy with respect to healthy individuals.

Types of Nanosensors

Electrical nanosensor   Electrical nanosensors incorporating nanostructures as sensing probes are highly promising devices for disease diagnostics in medicine. Chemical or biomolecular binding events are directly converted into electrical signals, which allows for label-free read-out in the case of DNA detection.

Electrochemical biosensor   Nanomaterials and nanotechnology bring new possibilities for biosensor construction and for developing novel electrochemical bioassays. The use of nanoscale materials for electrochemical biosensing has seen explosive growth in the past five years. Nanoscale materials have been used to achieve direct wiring of enzymes to electrode surface, to promote electrochemical reaction, to impose nanobarcode for biomaterials and to amplify signal of biorecognition event.

The electrochemical nanobiosensors are useful in cancer diagnostics and detection of infectious organisms. These devices are expected to develop reliable point-of-care in diagnosis of cancer and other diseases, and also as tools for intra-operative pathological testing, proteomics and systems biology. In-body electronic nanosensors may aid understanding of dynamic processes in cells and may provide new treatments for diseases.

Electrochemical biosensor works by immobilization of biomolecules. It is useful in bioaffinity assays for determining DNA and its effectors and also in electrochemical measurement of blood glucose concentration.

Nanowire biosensor   Nanowire-based biosensor arrays have significant impact for detection of biological threats, early diagnosis of cancer, drug discovery, and medical treatment. The nanowire-based biosensor arrays enable simultaneous detection of multiple analytes such as cancer biomarkers in a single chip, as well as fundamental kinetic studies for biomolecular reactions.

Viral nanosensor   Virus particles are essentially biological nanoparticles. Herpes simplex virus (HSV) and adenovirus have been used to trigger the assembly of magnetic nanobeads as a nanosensor for clinically relevant viruses. This system is more sensitive than ELISA-based methods and is an improvement over PCR-based detection because it is cheaper, faster and has fewer artefacts.

Nanoshell biosensors   Gold nanoshells have been used in a rapid immunoassay capable of detecting analyte within complex biological media without any sample preparation. Aggregation of antibody/nanoshell conjugates with extinction spectra in the near infrared is monitored spectroscopically in the presence of analyte. Nanoshells are already being developed for applications including cancer diagnosis, cancer therapy, and diagnosis and testing for proteins associated with Alzheimer's disease.

Nanotube-based biosensors   This is a biosensor based on single-wall carbon nanotube (SWNT) for the detection of biomolecules. By coating the surfaces of tiny carbon nanotubes with monoclonal antibodies, cancer cells in a tiny drop of blood can be detected. This approach is used towards developing a biosensor for breast cancer detection, by functionalizing the CNTs with antibodies that are specific to cell-surface receptors of breast cancer cells.

A system for biomolecular assays that uses carbon nanotubes as both an electrode and an immobilization phase in an electrochemiluminescence-based sensing device is developed. Detection of biomolecules like Streptavidin and IgG, application in immunoassays, nucleic acid probe assays, assays for clinical chemistry analytes are the uses of these biomarkers.

Applications of Nanobiosensors

The potential applications of nanobiosensors, the latest generation of electronic noses are numerous and promising and will replace chemical sensors that are only able to detect a single substance. Some areas where nanobiosensors are useful are:

1. food safety and quality control (toxic compounds, degradation, purity, rotten food)

2. testing of cosmetics and pharmaceuticals

3. protection of the environment (air and water quality, contaminants, smoke analysis)

4. on-line monitoring of industrial or environmental processes (ripening, fermentation)

5. safety controls (detection of dangerous or toxic substances, explosives at airports, drugs)

6. as an aid when searching for buried victims

7. medical diagnostics (diabetes, schizophrenia, cancer, etc.).

NANOBIOSENSORS AND CANCER

Various biosensor applications for cancer diagnostics are well known. Nanobiosensor plays very important role in cancer care. Bio-conjugated particles and devices are under progress for early cancer detection in body fluids such as blood and serum. These nanoscale devices operate on the principles of selectively capturing cancer cells or target proteins. The sensors are often coated with a cancer-specific antibody or other biorecognition ligands so that the capture of a cancer cell or target protein yields an electrical, mechanical, or optical signal for detection. New nanosensor uses quantum dots to detect DNA. Using tiny semiconductor crystals, biological probes and a laser, a new method of finding specific sequences of DNA by making them light up beneath a microscope are developed. Detecting a sample of DNA containing a mutation linked to ovarian cancer is possible by this method. A tiny biosensor to detect cancer proteins and potentially the bacteria that causes MRSA (methicillin-resistant *Staphylococcus aureus*) is under development. The device would work by identifying cancer markers— proteins or other molecules produced by cancer cells. These vary according to the type of cancer and are distinct from proteins produced by healthy cells.

The nanobiosensor has many other applications for looking at how cells react when they are treated with a drug or invaded by a biological pathogen. This has important implications ranging from drug therapy development to national security, environmental protection and a better understanding of molecular biology at a systems level.

Point-of-Care Testing

One of the applications of biosensors is for point-of-care testing (POCT), i.e., diagnostic testing performed on site and it offers the potential for faster and cheaper diagnostics.

The basic components required for using biosensors in clinical cancer diagnosis are shown in the Figure 7.10.

Cancer biomarkers identified from basic and clinical research, and from genomic and proteomic analyses must be validated. Ligands and probes for these markers can then be combined with detectors to produce biosensors for cancer-related clinical testing. Point-of-care cancer testing requires integration and automation of the technology as well as development of suitable sample preparation methods. Point- of-care testing is beneficial for cancer care but at the same time many challenges like development of reproducible biomarker assays; development of multi-channel biosensors; advances in sample preparation and cncer cell enrichment; development of new, more sensitive transducers, advanced manufacturing techniques, etc. are still to be overcome.

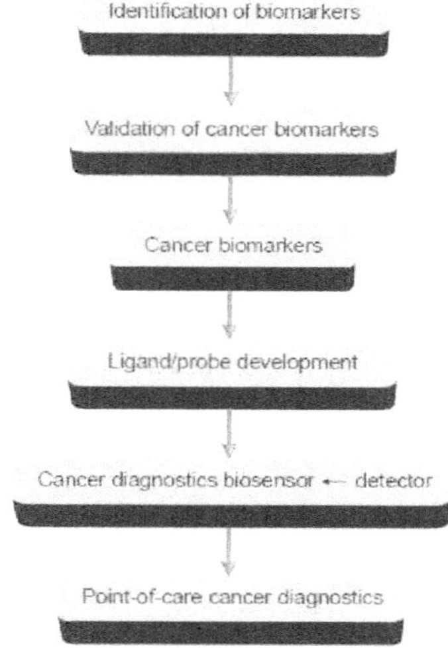

Figure 7.10   Clinical cancer diagnosis

NANO- DNA TECHNOLOGY

DNA nanotechnology is a sub branch of nanotechnology which uses the unique molecular recognition properties of DNA and other nucleic acids to create novel, controllable structures out of DNA. The DNA in this scientific work-up is used as a structural material and not as a carrier of genetic information, a form of an example of bionanotechnology. This has possible applications in molecular self-assembly and in DNA computing.

DNA nanotechnology makes use of branched DNA structures to create DNA complexes with useful properties. DNA is normally a linear molecule, in that its axis is unbranched. However, DNA molecules containing junctions can also be made. For example, a four-arm junction (Figure 7.11) can be made using four individual DNA strands which are complementary to each other in the correct pattern. Due to Watson–Crick base-pairing, only portions of the strands which are compmentary to each other will attach to each other to form duplex DNA. This four-arm junction is an immobile form of a Holliday junction.

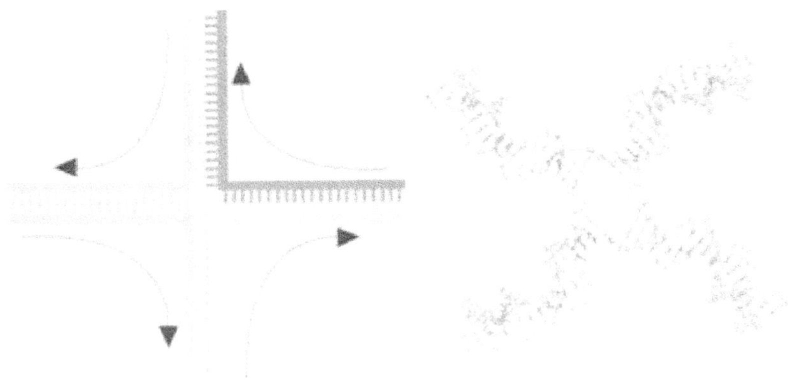

Figure 7.11   Structure of the four-arm junction

Junctions can be used in more complex molecules. The most important of these is the "double-crossover" or DX motif (Figure 7.12). Here, two antiparallel DNA duplexes lie next to each other and share two junction points where strands cross from one duplex into the other. This molecule has the advantage that the junction points are now constrained to a single orientation as opposed o being flexible as in the four-arm junction. This makes the DX motif suitable as a structural building block for larger DNA complexes.

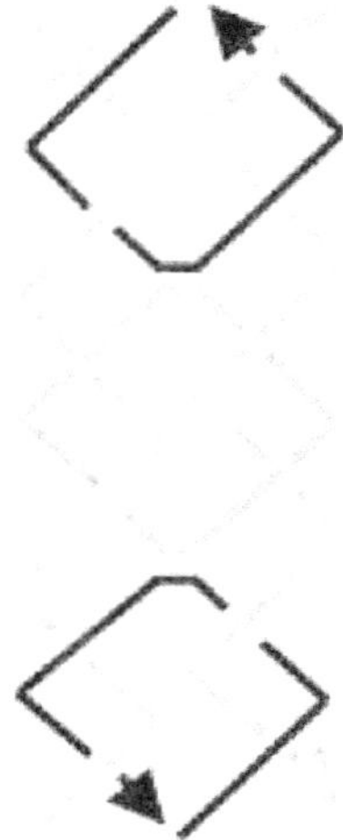

Figure 7.12   A double-crossover (DX) molecule (Mao, 2004)

TILE-BASED ARRAYS

DX arrays   DX molecules can be equipped with sticky ends in order to combine them into a two-dimenstional periodic lattice. Each DX molecule has four termini (Figure 7.13), one at each end of the two double-helical domains and these can be equipped with sticky ends that program them to combine into a specific pattern. More than one type of DX can be used which can be made to arrane in rows or any other tessellated pattern. They thus form extended flat sheets which are essentially two-dimensional crystals of DNA.

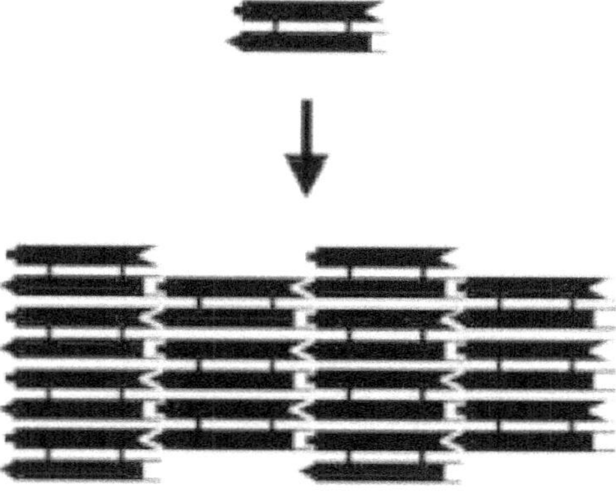

Figure 7.13   Assembly of a DX array (Mao, 2004)

DNA nanotubes   In addition to flat sheets, DX arrays have been made to form hollow tubes of 4–20 nm diameter. These DNA nanotubes are somewhat similar in size and shape to carbon nanotubes, but the carbon nanotubes are stronger and better conductors, whereas the DNA nanotubes are more easily modified and connected to other structures.

Other tile arrays   Two-dimensional arrays have been made out of other motifs as well, including the Holliday junction rhombus array as well as various DX-based arrays in the shapes of triangles and hexagons. Another motif, the six-helix bundle, has the ability to form three-dimensional DNA arrays as well.

DNA origami   As an alternative to the tile-based approach, two-dimensional DNA structures can be made from a single, long DNA strand of arbitrary sequence which is folded into the desired shape by using shorter, "staple" strands. This allows the creation of two-dimensional shapes at the nanoscale using DNA. Designs created out of this method include the smiley face and a coarse map of North America.

DNA polyhedra   A number of three-dimensional DNA molecules have been made which have the connectivity of a polyhedron such as an octahedron or cube. In other words, the DNA duplexes trace the edges of a polyhedron with a DNA junction at each vertex. The earliest demonstrations of DNA polyhedra involved multiple ligations and solid-phase synthesis steps to create catenated polyhedra. A DNA truncated octahedron made from a long single strand as well as a tetrahedron which can be produced from four DNA strands in a single step are also made.

DNA nanomechanical devices   DNA complexes have been made which change their conformation upon some stimulus. These are intended to have applications in nanorobotics. One of the first such devices, called "molecular tweezers," changes from an open to a closed state based upon the presence of control strands.

DNA machines have also been made which show a twisting motion. One of these makes use of the transition between the B-DNA and Z-DNA forms to respond to a change in buffer conditions. Another relies on the presence of control strands to switch from a paranemic-crossover (PX) conformation to a double-junction (JX2) conformation.

APPLICATIONS

DNA nanotechnology has been applied to the related field of DNA computing. A DX array has been demonstrated whose assembly encodes an XOR operation, which allows the DNA array to implement a cellular automaton which generates a fractal called the Sierpinski gasket (Figure 7.14). This shows that computation can be incorporated into the assembly of DNA arrays, increasing its scope beyond simple periodic arrays.

DNA computing overlaps with, but is distinct from, DNA nanotechnology. The latter uses the specificity of Watson–Crick base-pairing to make novel structures out of DNA. These structures anbe used for DNA computing. DNA computing can be done without using the types of molecules made possible by DNA nanotechnology also.

Figure 7.14   DNA arrays that display a representation of the Sierpinski gasket on their surfaces (Rothemund *et al.*, 2004)

Nano-architecture   The idea of using DNA arrays to template the assembly of other functional molecules has been around for a while, but only recently has progress been made in reducing these kinds of schemes to practice. In 2006, researchers convalently attached gold nanoparticles to a DX-based tile and showed that self-assembly of the DNA structures also assembled the nanoparticles hosted on them. The first report of assembly with a non-covalent hosting scheme came in 2007, using Dervan polyamides on a DX array to arrange streptavidin proteins on the DNA array.

Building Blocks of DNA

1. Every known living thing uses DNA to store its genetic code. It is this code that determines everything about the organism. The code is like a cookbook that contains instructions on how to construct a biological entity.

2. The DNA molecule can be divided into sections called genes that determine how to genetically build something. For example, a subset of DNA that is 100,000 base pairs long could describe how to make lactase, an enzyme that digests milk in the small intestine. People who are lactose-intolerant are not producing the enzyme because their bodies are not reading that piece of DNA correctly.

3. A DNA molecule consists of two long chains of amino acids that twist around each other in a structure called a double helix (Watson and Crick model).[2] There are four types of nucleic acids, which form complementary pairs: adenine pairs with guanine; thymine pairs with cytosine. One chain of the double helix is the

complement of the other chain, so the information in a DNA molecule is stored twice. This allows copies to be made from one side of the helix.

4. Scientists often describe a short length of DNA by just listing abbreviations for its amino acids—AATCGTGCTA. A single letter, which actually represents an amino acid and its complement, is called a base pair. An entire DNA molecule for a human consists of about 3 billion base pairs. That would make almost 200,000 newsprint pages filled with only AGTCGTCG...

5. Identical copies of the DNA are stored in most cells of the organism (unique cells for reproduction, namely sperm and eggs, are an exception). When a cell divides in normal growth, it makes a copy of the DNA it contains and places a copy into each new cell.

6. A cancer is a growth of cells whose DNA has become corrupted. Cancers can be very devastating because their corrupted DNA causes the cell to reproduce far more quickly than normal. This leads to malignant tumours that can impede the function of a body's organs.

The discovery of the polymerase chain reaction (PCR) created the way to a new era of biological research and applications. The impact can be felt in the field of molecular biology, as well as other allied fields of science including medicine. Novel classes of semi-synthetic DNA-protein conjugates, self-assembled oligomeric networks consisting of streptavidin and double-stranded DNA, which can be converted into well-defined supramolecular nanocircles, have been developed.

The DNA-streptavidin conjugates are applicable as modular building blocks for the production of new immunological reagents for the ultrasensitive trace analysis of proteins and other antigens by means of immuno-PCR methodology. ImmunoPCR is a combination of the specificity of an antibody-based immunoassay with the exponential power of the amplification of PCR, hence resulting in a 1000-fold degree of sensitivity as compared with standard ELISA (Enzyme-linked immunosorbent assay) methods. Piezoelectric biosensor is based on nucleic acid interaction and is DNA-based. It is useful in point mutation detection in PCR samples.

Self-assembled DNA-streptavidin conjugates have also been applied in the field of nanotechnology such as the following.

1. The conjugates are used as model systems for ion-switchable nanoparticle networks, as nanometre-scale "soft material" calibration standards for scanning probe microscopy.

2. They are used as programmed building blocks for the rational construction of complex biomolecular architecture, which may be used as templates for the growth of nanometre-scale inorganic devices.

3. Covalent conjugates of single-stranded DNA and streptavidin are used as biomolecular adapters for the immobilization of biotinylated macromolecules at solid substrates through nucleic acid hybridization. This "DNA-directed immobilization" allows for reversible and site-selective functionalization of solid substrates with metal and semiconductor nanoparticles or, vice versa, for the DNA-directed functionalization of gold nanoparticles with proteins, such as immunoglobulins and enzymes. The fabrication of functional biometallic nanostructures from gold nanoparticles and antibodies are applied as diagnostic tools in bioanalytics.

DNA SENSORS

Sensors which utilize the DNA molecules for applications in medicine and allied areas are called as DNA sensors. DNA sensors are put into use to detect diseases, to warn biochemical attacks or test meant for contamination and also for biological identification. A handy device that can identify a person infected with HIV, in seconds with a drop of their blood, is a possibility by new technology of DNA sensor. The way to sense specific, short segments of DNA gives rise to very powerful and beneficial devices.

The DNA sensor (Figure 7.15) detects the presence of small amounts of a specific segment of DNA by changing an electrical signal. The principle of detection is using nucleic acid hybridization rennealing between the ssDNAs from different sources. Each sensor must be custom-made for the detection of a segment of DNA. For example, a sensor may be manufactured to sense HIV DNA, whie another may sense a biochemical attack. A device that could sit in subway stations and sniff the air looking for pathogens is a possibility soon.

The sensor consists of a gold electrode that is coated with special loop-shaped DNA molecules. Each molecule can assume two shapes. In the looped shape, the tail of the molecule is held close to the gold surface. In the stretched shape, the tail is held further away from the gold. The molecule changes shape when it finds and bonds with its complement segment.

The tail of the molecule contains an electrochemical agent that allows an electric current to flow when it is close to the gold electrode, but restricts the current when pulled away from the gold. The tail moves a little distance (a few nanometres) in such a situation. The electric current stops flowing when the molecule changes shape. It is then amplified and displayed by an electric circuit.

The innovation is, these DNA sensors do not need any separate chemicals or reagents (reagentless and reusable) to do the test.

Figure 7.15   A DNA testing apparatus

To return to their looped state, DNA sensor molecules need the following procedure. By pouring hot water, DNA molecules are joined with hydrogen bonds, which are sensitive to heat. Washing the gold electrode with hot water will wash away the bonded molecules and ready the sensor for another test.

Some of the issues that affect sensitivity may be dirt, which covers the sensor and clogs it, and other chemicals that partially damage the sensor. Changing the sensor in such a way that its form is less susceptible to dirt and more inert may solve the problems.

Once the sensors are able to detect pathogens in normal blood samples, a medical revolution will be underway to build sensors that can detect any choice of diseases. It is entirely possible to build a sensor that could detect a dozen diseases at one time. Placing an assortment of sensor molecules on the same electrode could do this. Such a system would be useful for screening patients against diseases rapidly. DNA biosensors can detect the presence of genes or mutant genes associated with inherited human diseases.

DNA Field-effect Transistor (DNAFET)

A DNA field-effect transistor (DNAFET) is a field-effect transistor which uses the field-effect due to the partial charges of DNA molecules to function as a biosensor. The structure of DNAFETs is similar to that of MOSFETs with the exception of the gate structure which, in DNAFETs, is replaced by a layer of immobilized ssDNA (single-stranded DNA) molecules which act as surface receptors. When complementary DNA strands hybridize to the receptors, the charge distribution near the surface changes, which in turn modulates current transport through the semiconductor transducer.

Arrays of DNAFETs can be used for detecting single nucleotide polymorphisms (causing many hereditary diseases) and for DNA sequencing. Their main advantage compared to optical detection methods in common use today, is that they do not require labelling of molecules. Furthermore they work continuously and (near) real-time. DNAFETs are highly selective since only specific binding modulates charge transport.

OPTICAL BIOSENSORS

A sensor that uses light to detect the effect of a chemical on a biological system is an optical biosensor. Optical biosensors are the ones using optical transducers in which the output measured is light intensity. The construction and use of optical nanosensors was first reported by Kopelman in 1992. This is a form of transducer that converts the biochemical interactions into a measurable electronic signal namely light intensity. Optical nanosensors, like larger sensors, can typically be classified into one of two wide categories, chemical or biological, depending on the probe used. Both types of sensors have been used to offer a reliable method of monitoring various chemicals in microscopic environments and have even been used to detect different entities within single cells.

There are two main areas of development in optical biosensors. They are:

1. determining changes in light absorption between the reactants and products of a reaction

2. measuring the light output by a luminescent process.

The former involves the widely established and utilized colorimetric test strips. They are disposable single-use cellulose pads impregnated with enzyme and reagents. The most common use of this technology is for whole-blood monitoring in diabetes control. In this case, the strips include glucose oxidase, horseradish peroxidase and a chromogen (e.g. *o*-toluidine or 3,3´,5,5´-tetramethylbenzidine). The hydrogen peroxide, produced by the aerobic oxidation of glucose (see reaction scheme), oxidizes the weakly coloured chromogen to a highly coloured dye.

$$\text{Glucose} + \text{Peroxidase} = H_2O_2 \ (\text{Hydrogen peroxide})$$
$$\text{Chromogen} \ (2H) + H_2O_2 = \text{Dye} + 2H_2O$$

The evaluation of the dyed strips is best done by the use of portable reflectance meters. Direct visual comparison with a coloured chart is also used. A wide variety of test strips involving other enzymes are commercially available now.

The most promising biosensor involving luminescence uses firefly luciferase [*Photinus*-luciferin 4-monooxygenase (ATP-hydrolysing)], to detect the presence of bacteria in food or clinical samples.

Nanosized Optical Biosensors

Tiny specs of silver could play a central role in a sensitive and inexpensive biological detection system. The nanoparticle-based analytical procedure might be used to monitor ligand–receptor binding events, protein adsorption on self-assembled monolayers, and other types of analyte–surface interactions.

Based on a technique in which researchers measure changes in the extinction (absorption plus scattering) spectrum of microscopic metal particles, the new procedure provides a way to measure low concentrations of specific analyte molecules using simple UV-Visible spectroscopy instrumentation.

The detection principle is based on a variation of surface plasmon resonance (SPR) spectroscopy. In this method, centimetre-sized, thin gold films functionalized with receptor molecules serve as substrates for detection. Binding molecules to the surface causes a change in the system's index of refraction that in turn causes a shift in the SPR wavelength.

Nanosized silver triangles—actually flat-topped pyramids—as substrates are used. By switching to the nanometre scale, the substrate's spectral properties are sensitive to the particles size and shape. By using two-dimensional arrays of nanoparticles, the method's lateral resolution, surface sensitivity, and other properties are improved compared to traditional forms of SPR spectroscopy.

When light of arbitrary wavelength is shone on nanoparticles, most of the light is reflected. But, at specific wavelengths the particles absorb light. In the case of the silver nanoparticles, absorption is due to a collective excitation of nearly 10 million silver atoms in each particle.

Biosensors Based on Localized Surface Plasmon Resonance Spectroscopy

The Ag-nanoparticle-based localized surface plasmon resonance (LSPR) nanosensor (Figure 7.16) yields ultrasensitive biodetection with extremely simple, small, light, robust, and low-cost instrumentation. Using LSPR spectroscopy, the model system, biotinylated surface confined Ag nanotriangles, was used to detect less than one picomolar up to micromolar concentrations of streptavidin. Additionally, the monitoring of anti-biotin binding to biotinylated Ag nanotriangles exhibited that the system could be used as a solution immunoassay. The system was rigorously tested for nonspecific binding interactions and was found to display virtually no adverse results. These results repesent important new steps in the development of the LSPR nanobiosensor for applications in medical diagnostics, biomedical research, and environmental science.

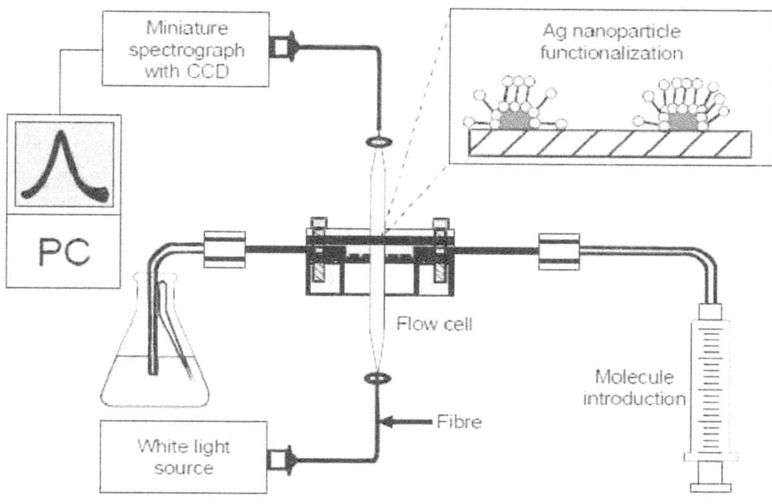

Figure 7.16   LSPR nanosensor experiment

The flow cell is fibre-optically coupled to a white light source and miniature spectrometer. The cell is directly linked to either a solvent reservoir or to a syringe containing the desired analyte.

Ag nanotriangle biosensors could be used for the detection of a wide variety of biomolecules. Binding of DNA, proteins, and possibly eukaryotic cells (by using protein– ligand intermediates) to noble metal nanoparticles opens a window of opportunity in medical diagnostics and could greatly simplify often tedious immunohistochemical detection tasks performed regularly in biomedical research. Future work on miniaturization of the sensor, linkage of the sensor to drug delivery chips, and biocompatibility could make this laboratory-based device into a portable diagnostic tool. The simplicity, durability, and reusability of the LSPR nanobiosensor has potential use as an affordable and efficient medical device. It is useful for the detection and identification of biological analytes and biophysical analysis of biomolecular interactions.

Supramolecular Assembly

It is a newer addition to the biological research and application in medical fields like tissue engineering, regenerative medicine, and targeted drug delivery. A well defined complex of molecules held together by noncovalent bonds[3] is called a supramolecular assembly or "supermolecule". A supramolecular assembly can be composed of two molecules like a DNA double helix or an inclusion compound. It denotes larger complexes of molecules that form sphere-, rod-, or sheet-like species. The dimensions of supramolecular assemblies can range from nanometres to micrometres. Thus they allow access to nanoscale objects using a bottom-up approach in far fewer steps than a single molecule of similar dimensions.

Molecular self-assembly is the process by which a supramolecular assembly is formed. Self-assembly is the process by which individual molecules form the defined aggregate. Self-organization, then, is the process by which those aggregates create higher-order structures. These are useful related to liquid crystals and block copolymers. Figure 7.17 shows examples of supramolecular assemblies.

Applications Supramolecular assemblies are considered as new materials in a variety of contexts. A supramolecular assembly of peptide amphiphiles[4] in the form of nanofibres could be used to promote the growth of neurons. An important advantage to this supramolecular approach is that the nanofibres will degrade back into the individual peptide molecules that can be broken down by the body.

Self-assembling dendritic dipeptides, which form hollow cylindrical supramolecular assemblies in solution and in bulk are having implications in the biology/materials science interface. The cylindrical assemblies possess internal helical order and self-organize into columnar liquid crystalline lattices. When inserted into vesicular membranes, the porous cylindrical assemblies mediate transport of protons across the membrane.

Self-assembling dendrons have been used to generate arrays of nanowires. Electron donor–acceptor complexes comprise the core of the cylindrical supramolecuar assemblies (Figure 7.17). Each cylindrical supramolecular assembly functions as an individual wire. High-charge carrier mobilities for holes and electrons are obtained.

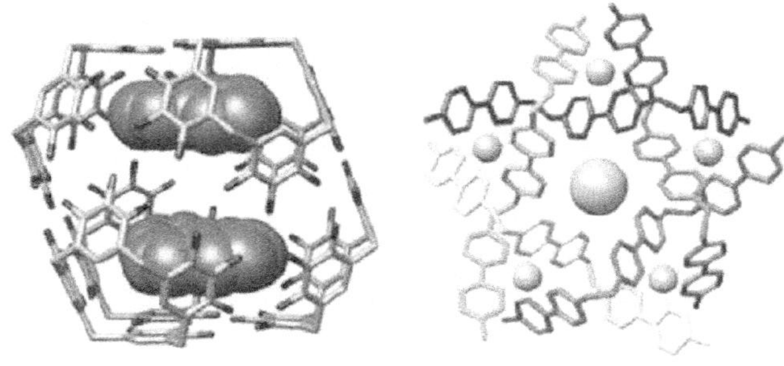

Figure 7.17 Supramolecular assembly

NANOPATHOLOGY

Disease-producing agents like bacteria, viruses and parasites, can be the cause of various pathologies of human body is a well-known fact. In clinical medicine, it is a well known fact that inhaled particles can induce diseases like asbestosis and silicosis. Another well-known clinical phenomenon is that a number of implantable medical devices wear *in-vivo*, thus creating debris of micro- and nanosized particulate matter. Examples of this are the wear of hip-joint prostheses and of dental restorations. Inorganic, chemically inert, microscopic debris can induce granulomatoses even in regions beyond the implant site is a possibility and familiar to orthopaedic surgeons. So, they must remove worn hip-joint prostheses because the debris derived out of their erosion, brings about the local formation of granulomatous tissue and a bone degeneration with the ensuing loosening of the device. Till recently an important area of pathological agents for human diseases is very much ignored related to the foreign bodies below a certain size, which comes from environmental pollution mostly. They can enter animal organisms including humans, mainly through ingestion or inhalation and travel more freely through their tissues. They can cross through the gastro-intestinal wall or the pulmonary alveoli and be carried by the blood or the lymph in which they enter or occupy a tissue as they move and on their migratory way. As per their size, these foreign bodies can be classified into nanoparticles, if their size is between 10^{-9} and 10^{-7} m and microparticles if it is between 10^{-6} and 10^{-5} m. Foreign bodies if chemically inert and non-biodegradable, then a reaction develops through which the organism defends itself against that form of invasion. This reaction, whatever it is, can either pass unnoticed or be of clinical relevance.

Nanopathology is the sub-branch of pathology which deals with the reactions by the organism to the presence of those small-sized nano- or microparticles.

Interactions

Pathology can be started by the presence of inorganic particles that cannot be metabolized or disposed of from the body. The size of the particles, their local concentration and their velocity to reach the critical concentration can have an influence on the type of pathology. Because of the above information, the concept of biocompatibility should be revised. A material, which is certainly accepted in bulk form, may be no longer biocompatible when its size is reduced below a certain "critical" threshold. The different chemistry of the particles (either ceramic or metal or plastic materials) can influence the relationship between the cells and the material's surface, which can lead to a different cellular reaction. This may lead to a clinical symptom or sign but no literature is available.

The biological activity of a tissue in contact with a foreign body differs depending both on the tissue and on the foreign body's chemistry, surface, shape and size. It is very difficult to define a material "biocompatible". Such definition takes the material's chemistry and its surface, but rarely the shape that particular material takes when it is transformed into an implantable device. Something that comes in contact with a living tissue and rarely its size are taken into consideration while calling it as biocompatible.

In order to be biocompatible, a foreign body that is put in contact with a biological tissue must induce a specific protein adsorption from the extracellular matrix. There are many cellular processes, which are triggered by the type of protein adsorbed, by its conformation and by its biological activity. A certain protein is needed to guarantee a proper interaction of the foreign body like an implantable medical device with the biological environment is evolved as an observation. It is also projected that manipulating the implant surface in order to induce in advance that situation is also practical. The "biomimetic" surfaces which are used now based their activity in the human body upon this concept. Biotechnologies and nanotechnologies support the ability to design such a system. Medical applications of nanotechnologies are already available, for instance in dentistry, by the creation of nanocomposites for dental restorations.

At present, dental restorations employ amalgams, composites, or gold or ceramic inlays. But there is an urgent need to have more resistant materials, especially for posterior teeth like molar and premolar, where the mastication load is particularly high. The materials capable of replacing the amalgam, long suspected to be toxic, is the goal due to overload. Some dental materials employing nanoparticles, for instance of zirconia, are already available on the market. More sophisticated, materials are being developed and are on the pipeline. The following points are pertinent 1) These materials interact in different ways with the endothelium, macrophages and gut and liver epithelial cells. 2) Metals are more dangerous than plastics for the survival activity of the cells. 3) The production of the inflammtory and defence mediators depends on the particle chemistry. *In vitro* tests give the idea that new phenomena can occur when the nanoscale range of interaction is investigated.

ENDNOTES

1. fmol—symbol for femonsle (an SI unit of amount of hy equal to 10^{-15}) mole.

2. Elucidation of structure of **DNA** by **James Watson and Francis Crick** Studies of DNA had revealed much about its chemical and physical nature, but Watson believed that its function could not be understood fully until its structure was known. Crick and Watson used the results from previous studies and X-ray diffraction data from Maurice Wilkins and Rosalind Franklin to help them determine DNA's molecular structure. By 1953 they had built a model which incorporated all known features of DNA, and proposed the double helix

structure which is commonly referred to as the Watson-Crick Model of DNA. Crick, Watson and Wilkins were awarded Nobel prize for this work in 1962.

3. A noncovalent bond is a type of chemical bond, typically between macromolecules, that does not involve the sharing of pairs of electrons, but rather involves more dispersed variations of electromagnetic interactions. The noncovalent bond is the dominant type of bond between supermolecules in supermolecular chemistry. Noncovalent bonds are critical in maintaining the three-dimensional structure of large molecules, such as proteins and nucleic acids and are involved in many biological processes in which large molecules bind specifically but transiently to one another.

4. Peptide-amphiphiles are amphiphilic structures with a hydrophilic peptide head group that incorporates a bioactive sequence and has the potential to form distinct structures, and a hydrophobic tail that serves to align the head group, drive self-assembly, and induce secondary and tertiary conformations. Peptide-amphiphiles have found applications as soft bioactive materials for mdel studies of bioadhesion and characterization of different cellular phenomena, as well as scaffolds for tissue engineering, regenerative medicine, and targeted drug delivery.

REVIEW QUESTIONS

1. Discuss the basic concepts of nanomedicine and its applications.

2. What is nanobiotechnology and nanobiotechnological devices.

4. What are nanoshells?

5. Give an account of biosensors and nanobiosensors.

7. What are the types of nanosensors?

8. Discuss nanobiosensors in relation to cancer.

9. What are DNA sensors?

10. Discuss in detail about optical biosensors.

11. Write a note on nanopathology.

12. Write short notes on:

    i.   Nanosized optical biosensors

    ii. Surface plasmon resonance spectroscopy

    iii. Drug delivery

iv.   Cancer diagnosis and therapy

v.   *In vivo* therapy

vi.   Neuro-electronic interfaces

vii.   Cell repair machines

viii.   Fluorescent biological labels

ix.   Colorimetric assay

x.   Drug delivery system

xi.   Dendrimers

xii.   Nanorobots

xiii.   Supramolecular chemistry

xiv.   Nanonose

xv.   Applications of nanobiosensors.

xvi.   Point-of-care testing

xvii.   Nano-DNA-technology

xviii.   DNA field-effect transistor (DNAFET)

REFERENCES

1.   Allen, T.M. and Cullis, P.R. (2004). "Drug delivery systems: entering the mainstream." Science. 303 (5665): 1818–1822.

2.   Foster, L.E. (2006). *Medical Nanotechnology*: Science, Innovation, and Opportunity. Pearson Education, Upper Saddle River.

3.   Freitas, A. Robert, Jr. (2005). "Current status of nanomedicine and medical nanorobotics." *Journal of Computational and Theoretical Nanoscience.* 2: 1–25.

4.   Freitas, R.A. Jr. (2005). "What is Nanomedicine?." *Nanomedicine: Nanotech. Biol. Med.* 1 (1): 2–9.

5.   International Union of Pure and Applied Chemistry. "Biosensor." *Compendium of Chemical Terminology.* Internet edition.

6.   Loo, C., Lin, A., Hirsch, L. Lee, M.H., Barton, J., Halas, N., West, J. and Drezek, R. (2004). "Nanoshell-enabled photonics-based imaging and therapy of cancer." *Technology in Cancer Research and Treatment.* 3 (1): 33–40.

7.   Nie, Shuming, Yun Xing, Gloria, J. Kim and Jonathan, W. Simmons. "Nanotechnology applications in cancer." *Annual Review of Biomedical Engineering.* 9.

8. Ratner, M. and Ratner, D. (2002). *Nanotechnology: A Gentle Introduction to the Next Big Idea*. Pearson Education, Inc.

9. Robert, A. Freitas Jr. (1999). *Nanomedicine, Volume I: Basic Capabilities*. Landes Bioscience.

10. Robert, A. Freitas, Jr. (2003). *Nanomedicine, Volume IIA: Biocompatibility*. Landes Bioscience.

11. Coombs, R.R.H. and Robinson, D.W. (eds.). (1996). *Nanotechnology in Medicine and the Biosciences*. Gordon and Breach Publishers.

12. Seeman, Nadrian, C. (1999). "DNA engineering and its application to nanotechnology." *Trends in Biotechnology*. 17 (11): 437–443.

13. Virgil Percec, Andrés, E. Dulcey, Vekatachalapathy, S.K.Balagarusamy, Yoshiko Miura, Jan Smidrkal, Mihai Peterca,Sami Nummelin, Ulrica Edlund, Steven, D. Hudson, Paul, A. Heiney, Hu Duan, Sergei N. Magonov and Sergei A. Vinogradov. (2004). Self-assembly of amphiphilic dendritic dipeptides into helical pores." Nature. 430: 764–768.

14. Shi, X., Wang, S., Meshinchi, S. Van Antwerp, M.E., Bi X Lee, I. and Baker, J.R. Jr. (2007). "Dendrimer-entrapped gold nanoparticles as a platform for cancer-cell targeting and imaging." Small 3 (7): 1245–1252.

Chapter 8

Instruments and Methodology

INTRODUCTION

In the 1910s nanoscience and nanotechnology's first tools to measure and make nanostructures were possible. With the discovery of electrons and neutrons, the field of nanotechnology came into existence. These discoveries helped the scientists to accept that matter can exist on a much smaller scale than what is thought of as small, and/or what was the smallest structure possible to be created at that time. This has created the curiosity for development of the nanostructures.

The first observations and measurements of nanoparticles were made by Zsigmondy. He made detailed study of gold sols and other nanomaterials of sizes down to 10 nm and less. Zsigmondy utilized the ultramicroscope which employed dark-field method for seeing particles with sizes much less than the wavelength of light.

During the 20th century, traditional techniques in Interface and Colloidal Science were developed for characterizing nanomaterials. These materials are *first-generation* passive nanomaterials. These methods include techniques for characterizing particle size distribution, which is necessary because many materials supposed to be nano-sized are actually aggregated in solutions. Some methods are based on light scattering and ultrasound attenuation spectroscopy for testing concentrated nano-dispersions and microemulsions.

There are traditional techniques for characterizing surface charge or zeta potential[1] of nanoparticles in solutions. This information is used for the proper system stabilization and preventing its aggregation or flocculation. These methods include microelectrophoresis, electrophoretic light scattering and electroacoustics. The last one, namely, colloid vibration current method is suitable for characterizing concentrated systems.

NEXT GENERATION OF NANOTECHNOLOGICAL TECHNIQUES

They include the following techniques which are enumerated.

1. Fabrication of nanowires
2. Semiconductor fabrication.
    i. deep ultraviolet lithography
    ii. electron beam lithography
    iii. focused ion beam machining
    iv. nanoimprint lithography
    v. atomic layer deposition
    vi. molecular vapour deposition
3. Molecular self-assembly techniques

Techniques such as Di-block copolymers are extensions in the development of scientific advancements which have preceded the nanotech era rather than techniques for the sole purpose of creating nanotechnology and for utilization in nanotechnology research.

MODERN DEVELOPMENTS

The atomic force microscope (AFM) and the scanning tunnelling microscope (STM) are two early versions of scanning probes that launched nanotechnology. There are types of scanning probe microscopy, development based on the scanning confocal microscope and the scanning acoustic microscope (SAM)—to see structures at the nanoscale. The tip of a scanning probe can be used to manipulate nanostructures and the process is called positional assembly. Feature-oriented scanning-positioning methodology is a promising way to implement nanomanipulations in automatic mode. This is a slow process due to low-scanning velocity of the microscope.

Methodologies which are useful and developed in nanosciences are as follows.

TECHNIQUES OF NANOLITHOGRAPHY

Lithography is a top-down fabrication technique where a bulk material is reduced in size to nanoscale pattern.

1. Dip-pen nanolithography
2. Electron beam lithography
3. Nanoimprint lithography

The top-down approach anticipates nanodevices which are built piece by piece in stages. Scanning probe microscopy is an important technique for both characterization and synthesis of nanomaterials. Atomic force microscopes and scanning tunnelling microscopes can be used to look at surfaces and to move atoms around. By designing different tips for these microscopes, they can be used for carving out structures on surfaces and to help to guide self-assembling structures. By using feature-oriented scanning-positioning approach, atoms can be moved around on a surface with scanning probe microscope techniques. At present they are only suitable for laboratory experimentation.

Bottom-up techniques build larger structures atom by atom or molecule by molecule. These techniques include chemical synthesis, self-assembly and positional assembly and molecular beam epitaxy[2] (MBE). Samples made by MBE were responsible and important to the discovery of the fractional quantum Hall Effect[3](1998 Nobel Prize in Physics). In the process MBE allows scientists to lay down atomically-precise layers of atoms and, build up complex structures. MBE is also widely used to make devices for the newly emerging field of spintronics and is important for the research on semiconductors.

Newer techniques like dual polarization, interferometry are useful in measuring quantitatively the molecular interactions that take place at the nanoscale.

SCANNING PROBE MICROSCOPE

Scanning probe microscope (SPM) is a type of microscope which is useful in the nanotechnological field. The basics of SPM were formed with the invention of the scanning tunnelling microscope in 1981. Images of surfaces using a physical probe that scans the specimen is formed in SPM. An image of the surface is obtained by mechanically moving the probe in a raster scan[4] of the specimen, line by line. The recording of the probe-surface interaction is done as a function of position. Several interactions can be imaged simultaneously by many scanning probe microscopes. 'A' mode is the name given to the way of using these interactions to obtain an image.

The resolution varies from technique to technique and some probe techniques reach atomic resolution. The ability of piezoelectric actuators to execute motions with a precision and accuracy at the atomic level or better on electronic command produces this resolution. This group of technique is called as "piezoelectric techniques". The data are typically obtained as a two-dimensional grid of data points, visualized in false colour as a computer image in SPM field.

Types of Scanning Probe Microscopy

The following types of SPM are developed and put into use and the types are only enumerated to get some idea of the quantum of development in this field related to nanotechnology.

◈   AFM—atomic force microscopy

- ◈ contact AFM
- ◈ non-contact AFM
- ◈ dynamic contact AFM
- ◈ BEEM—ballistic electron emission microscopy
- ◈ EFM—electrostatic force microscope
- ◈ ESTM—electrochemical scanning tunnelling microscope
- ◈ FMM—force modulation microscopy
- ◈ KPFM—Kelvin probe force microscopy
- ◈ MFM—magnetic force microscopy
- ◈ MRFM—magnetic resonance force microscopy
- ◈ NSOM—near-field scanning optical microscopy (or SNOM, scanning near-field optical microscopy)
- ◈ PFM—piezo force microscopy
- ◈ PSTM—photon scanning tunnelling microscopy which uses an optical tip to tunnel photons
- ◈ PTMS—photothermal microspectroscopy/microscopy
- ◈ SECM—scanning electrochemical microscopy
- ◈ SCM—scanning capacitance microscopy
- ◈ SGM—scanning gate microscopy
- ◈ SICM—scanning ion-conductance microscopy
- ◈ SPSM—spin polarized scanning tunnelling microscopy which uses a ferromagnetic tip to tunnel spin-polarized electrons into a magnetic sample
- ◈ SThM—scanning thermal microscopy
- ◈ STM—scanning tunnelling microscopy. Many other microscopy techniques have been developed based upon STM.
- ◈ STP—scanning tunnelling potentiometry, which measures electric potential across a surface
- ◈ SVM—scanning voltage microscopy
- ◈ SHPM—scanning Hall probe microscopy

Advantages

1. The resolution of the microscopes is limited by the size of the probe–sample interaction volume (i.e., point spread function), which can be as small as a few picometres. Because of this, the measurement of small differences in object height

up to picometre level is possible. Laterally the probe–sample interaction extends only across the atom tip or atoms involved in the interaction.

2. The interaction can be used to modify the sample to create small structures (nanolithography).

3. Unlike electron microscope methods, specimens do not require a partial vacuum but can be observed in air at standard temperature and pressure or while submerged in a liquid reaction vessel.

Disadvantages

1. The detailed shape of the scanning tip is sometimes difficult to determine. Its effect on the resulting data is particularly noticeable if the specimen varies greatly in height over lateral distances of 10 nm or less.

2. The scanning techniques are generally slower in acquiring images, due to the scanning process. Like all scanning techniques, the embedding of spatial information into a time sequence opens the door to uncertainties in metrology.

3. The maximum image size is generally smaller.

4. Scanning probe microscopy is often not useful for examining buried solid–solid or liquid–liquid interfaces.

ATOMIC FORCE MICROSCOPE

The atomic force microscope (AFM) is a very high-resolution type of scanning force microscope (SFM). AFM has a resolution of fractions of a nanometre, more than 1000 times better than the optical diffraction limit. In 1986 Binnig, Quate and Gerber invented the first atomic force microscope (AFM). AFM has revolutionized nanotechnological development. The applications of this instrument are in the areas of intensive research for the future development of nanoscience. The AFM is one of the foremost tools for imaging, measuring and manipulating matter at the nanoscale. The information in this instrument is gathered by "feeling" (touching) the surface of the sample with a mechanical probe rather than looking as in an ordinary microscope. The first atomic force microscope is shown in Figure 8.1.

Piezoelectric elements that are present in the AFM and that facilitate tiny but accurate and precise movemnts on electronic command enable very precise scanning.

Figure 8.1   The first atomic force microscope

Description

Atomic force microscope (Figure 8.2) consists of a microscale cantilever with a sharp tip (probe) at its end. That tip is used to scan the specimen surface. The tip is brought into proximity of a sample surface. That makes the forces between the tip and the sample leading to a deflection of the cantilever according to Hooke's law. The cantilever is made up of silicon or silicon nitrde with a tip radius of curvature in the nanometre range.

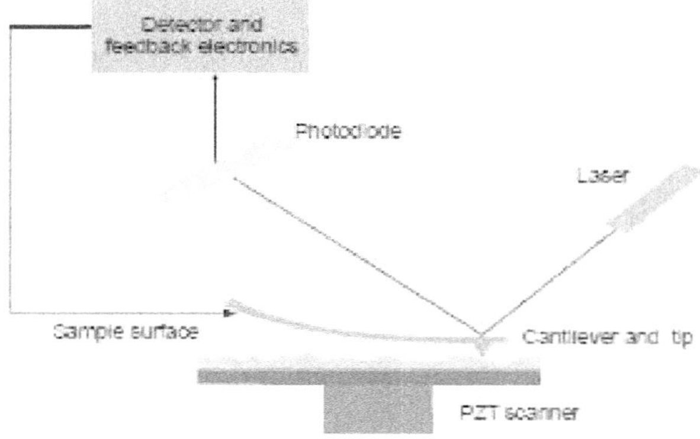

Figure 8.2   Functioning of atomic force microscope

A micro-fabricated cantilever with a sharp tip is deflected by features on a sample surface, much like in a phonograph but on a much smaller scale. A laser beam reflects off the backside of the cantilever into a set of photodetectors, allowing the deflection to be measured and assembled into an image of the surface.

Parameters Measured by AFM

Depending on the situation, the AFM measures the different forces. They are

1. Mechanical contact force

2. van der Waals forces

3. Capillary forces

4. Chemical bonding

5. Electrostatic forces

6. Magnetic forces as in magnetic force microscope (MFM)

7. Casimir forces

8. Solvation forces, etc.

Through the special type of probes like scanning thermal microscopy, photothermal microspectroscopy, etc. the additional information may be measured. The deflection is measured using a laser spot that reflected from the top of the cantilever into an array of photodiodes. Other methods are also used such as optical interferometry, capacitive sensing or piezoresistive AFM cantilevers. Due to deflection the strain in the AFM cantilever can be measured by using a Wheatstone bridge, but this method is not as sensitive as laser deflection or interferometry.

If the tip is scanned at a constant height, the risk of the tip colliding with the surface, causing damage is possible. A feedback mechanism is used to adjust the tip-to-sample distance. That mechanism will maintain a constant force between the tip and sample. This arrangement will prevent damage to the tip and surface. Usually, the sample is mounted on a piezoelectric tube that can move the sample in a direction for maintaining a constant force, and other directions for scanning the sample.

The AFM can be operated in a number of modes, depending on the application. "A mode" is the name given to the way of using these interactions to obtain an image. Imaging modes are divided into static or contact mode and a variety of dynamic or non-contact modes.

Imaging Modes

The primary modes of operation are the following.

1. *Static or contact mode*   In the static mode operation, the static tip deflection is used as a feedback signal. Because the measurement of a static signal is prone to noise and drift, low stiffness cantilevers are used to boost the deflection signal. However, close to the surface of the sample, attractive forces can be quite strong, causing the tip to "snap-in" to the surface. Thus static mode AFM is almost always done in contact

where the overall force is repulsive. Consequently, this technique is typically called "contact mode". In contact mode, the force between the tip and the surface is kept constant during scanning by maintaining a constant deflection.

2. *Dynamic non-contact mode*   In the dynamic mode, the cantilever is externally oscillated at or close to its fundamental resonance frequency or a harmonic. The oscillation amplitude, phase and resonance frequency are modified by tip–sample interaction force. These changes in oscillation with respect to the external reference oscillation provide information about the sample's characteristics.

Schemes for dynamic mode operation include:

i.   **Frequency modulation**   In frequency modulation, changes in the oscillation frequency provide information about tip–sample interactions. Frequency can be measured with very high sensitivity and thus the frequency modulation mode allows for the use of very stiff cantilevers. Stiff cantilevers provide stability very close to the surface and, as a result, this technique was the first AFM technique to provide true atomic resolution in ultra-high vacuum conditions.

ii. **Amplitude modulation**   In amplitude modulation, changes in the oscillation amplitude or phase provide the feedback signal for imaging. In amplitude modulation, changes in the phase of oscillation can be used to discriminate between the different types of materials on the surface. Amplitude modulation can be operated either in the non-contact or in the intermittent contact regime. In ambient conditions, most samples develop a liquid meniscus layer. Because of this, keeping the probe tip close enough to the sample for short-range forces to become detectable while preventing the tip from sticking to the surface presents a major hurdle for the non-contact dynamic mode in ambient conditions.

3. *Dynamic contact mode*   Also called intermittent contact or tapping mode, this mode was developed to bypass samples developing a liquid meniscus layer in ambient conditions. In dynamic contact mode, the cantilever is oscillated such that the separation distance between the cantilever tip and the sample surface is modulated.

Amplitude modulation has also been used in the non-contact regime to image with atomic resolution by using very stiff cantilevers and small amplitudes in an ultra-high vacuum environment.

4. *Tapping mode*   This isa form of single-molecule-under-water-AFM-tapping mode (Figure 8.3).

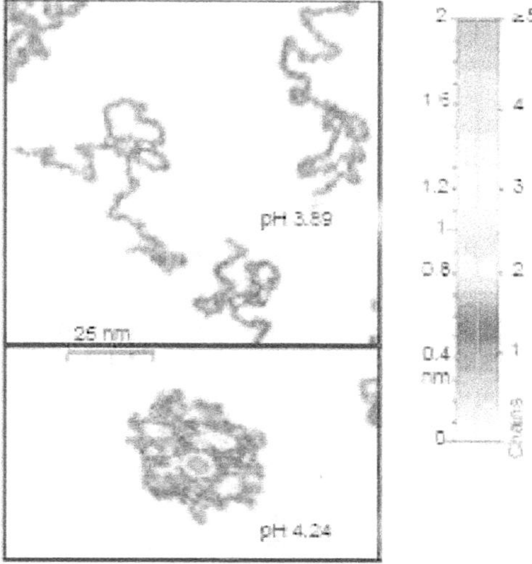

Figure 8.3   Single polymer chains (0.4 nm thick) recorded in a tapping mode under aqueous media with different pH (Roiter and Minko, 2005).

In tapping mode, the cantilever is driven to oscillate up and down at near its resonance frequency by a small piezoelectric element mounted in the AFM tip holder. The amplitude of this oscillation is greater than 10 nm, typically 100 to 200 nm. An electronic servo uses the piezoelectric actuator to control the height of the cantilever above the sample. The servo adjusts the height to maintain a set of cantilever oscillation amplitude as the cantilever is scanned over the sample.

A tapping AFM image is produced by imaging the force of the oscillating contacts of the tip with the sample surface. This is an improvement on conventional contact AFM, in which the cantilever just drags across the surface at constant force and can result in surface damage. Tapping mode is gentle enough even for the visualization of supported lipid bilayers or adsorbed single polymer molecules under liquid medium.

Force Spectroscopy

One of the major applications of AFM is force spectroscopy apart from imaging. Force spectroscopy is about the measurement of force–distance curves. These measurements are useful in measuring nanoscale contacts, atomic bonding, van der Waals forces, and Casimir forces, dissolution forces in liquids and single molecule stretching and rupture forces. Forces of the order of a few pico-newton can now be routinely measured with a vertical distance resolution of better than 0.1 nm. AFM tip is extended towards and retracted from the surface as the static deflection of the cantilever is monitored as a function of piezoelectric displacement.

Identification of Individual Surface Atoms

The AFM can be used to image and manipulate atoms and structures on a variety of surfaces. The atom at the apex of the tip "senses" individual atoms on the underlying surface when it forms incipient chemical bonds with each atom. Because these chemical interactions subtly alter the tip's vibration frequency, they can be detected and mapped.

These forces can be measured precisely for each type of atom and by this each different type of atom can be identified in the matrix as the tip is moved across the surface. The view of a sodium chloride using a AFM is shown in Figure 8.4.

This technique is used in biology, especially cell biology. Forces corresponding to (i) the unbinding of receptor–ligand couples (ii) unfoldig of proteins and (iii) cell adhesion at single cell scale are studied.

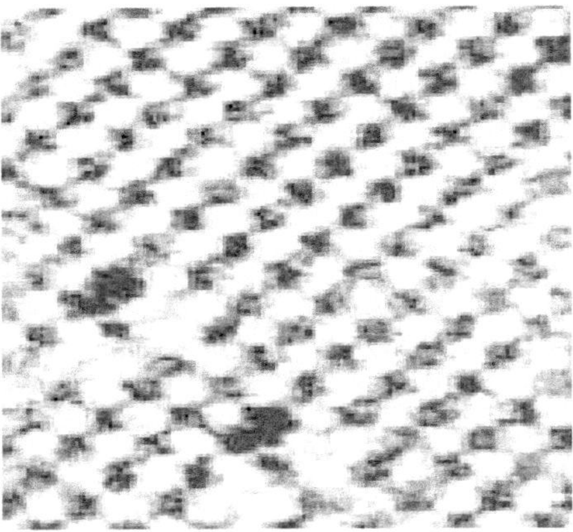

Figure 8.4   The atoms of a sodium chloride crystal viewed with an atomic force microscope

Advantages

The AFM has several advantages over the scanning electron microscope (SEM).

1.   True three-dimensional surface profile compared to two-dimensional profile of SEM.

2.   Samples do not require any special treatments (such as metal or carbon coatings) that would irreversibly change or damage the sample.

3.   AFM modes can work perfectly well in ambient air or even in liquid environment. This is useful to study biological macromolecules and even living organisms.

4. AFM can provide higher resolution than SEM–atomic resolution in ultra-high vacuum (UHV). UHV-AFM is comparable in resolution to scanning tunnelling microscopy and transmission electron microscopy (TEM).

Disadvantages

1. *Image size*   The SEM can image an area on the order of millimetres by millimetres with a depth of field on the order of millimetres. The AFM can only image a maximum height on the order of micrometres and a maximum scanning area of around 150 by 150 μm.

2. *Tip*   Incorrect choice of tip for the required resolution can lead to image artifacts.

3. *Speed*   AFM could not scan images as fast as an SEM, requiring several minutes for a typical scan, while a SEM is capable of scanning at near real-time after the chamber is evacuated. AFM is less suited for measuring accurate distances between artifacts on the image. Video AFM is a speedy variety. To eliminate image distortions induced by thermodrift, several methods are on.

4. *Images*   AFM images can be affected by hysteresis of the piezoelectric material and cross-talk between the (x,y,z) axes that may require software enhancement and filtering. Such filtering could "flatten" out real topographical features. Newer AFM use real-time correction software or closed-loop scanners which practically eliminate these problems. Some AFM also use separated orthogonal scanners (as opposed to a single tube) which also serves to eliminate cross-talk problems.

5. AFM cannot measure steep walls or overhangs. Specially made cantilevers can be modulated sideways as well as up and down (as with dynamic contact and non-contact modes) to measure sidewalls, at the cost of more expensive cantilevers and additional artifacts.

SCANNING TUNNELLING MICROSCOPE

Scanning tunnelling microscopy (STM) is a useful instrument in nanotechnological work and is also an important and powerful technique for viewing surfaces at the atomic level. In 1981, Gerd Binnig and Heinrich Rohrer developed this instrument and won the Nobel Prize in Physics (1986). STM probes the density of states of a material using tunnelling current. Scanning tunnelling microscopy (STM) is the starting point of scanning probe microscopy (SPM), a field of importance related to the development of nanotechnology. This has created the theoretical possibility to be practical and made the nanofield one of practical science. So, the credit of nanodevelopment goes to STM and is an important landmark in the nanotechnology field. For STM, good resolution is considered to be 0.1 nm lateral resolution and 0.01 nm depth resolution. The STM can be used not only in ultra-high vacuum but also in air and various other liquids or gases and at temperatures ranging from near 0 K to a few 100°C.

The STM is based on the concept of quantum tunnelling—when a conducting tip is brought very near to a metallic or semiconducting surface, a bias between the two can allow electrons to tunnel through the vacuum between them. For low voltages, this tunnelling current is a function of the local density of states (LDOS). Variations in current as the probe passes over the surface are translated into an image. STM is a difficult technique since it requires ultra-clean surfaces and sharp tips for experiment.

Tunnelling

Tunnelling is a concept from quantum mechanics. Classically, an object hitting an impenetrable wall of infinite height will bounce back. Throwing a baseball to the other side of a high brick wall directly at the wall is an example. Rather than bouncing back upon impact, the ball if simply passes through of the other side of the wall, is an observation related to tunnelling. For objects of very small mass, as is the electron, wavelike nature has a more pronounced effect, so such an event, referred to as tunnelling, has a much greater probability.

Instrumentation

The components of STM are:

1. Scanning tip—the radius of curvature of the scanning tip of the STM limits the resolution of an image. Image artifacts can occur if the tip has two tips at the end rather than a single atom leading to "double-tip imaging," a situation in which both tips contribute to tunnelling. It is essential to develop processes for consistently obtaining sharp, usable tips. The tip is often made of tungsten or platinum-iridium, though gold is also used. Tungsten tips are usually made by eletrochemical etching, and platinum-iridium tips by mechanical shearing (Figure 8.5).

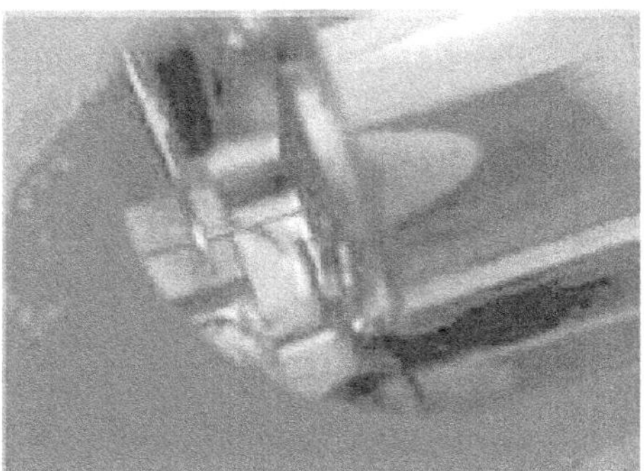

Figure 8.5   A close-up of a simple scanning tunnelling microscope head scanning MoS_2 using a platinum-iridium stylus

2. Piezoelectric controlled height

3. X, Y scanner

4. Coarse sample-to-tip control

5. Vibration isolation system—due to the extreme sensitivity of tunnel current to height, proper vibration isolation is imperative for obtaining usable results. Originally, magnetic levitation was used to keep the STM free from vibrations and now spring systems are used. Mechanisms for reducing eddy currents are also implemented.

6. Computer control   This is required for maintaining the tip position with respect to the sample, scanning the sample in raster fashion and acquiring the data. The computer is also used for enhancing the image with the help of image processing as well as performing quantitative morphological measurements.

Procedure

First the tip is brought into close proximity of the sample by some coarse sample-to-tip control. Once tunnelling is established, piezoelectric transducers are implemented to move the tip in three directions (Figure 8.6). As the tip is rastered across the sample in the X–Y plane, the density of states and therefore the tunnel current changes. This change in current with respect to position can be measured or the height of the tip corresponding to a constant current can be measured. Thesetwo modes are called constant height mode and constant current mode, respectively.

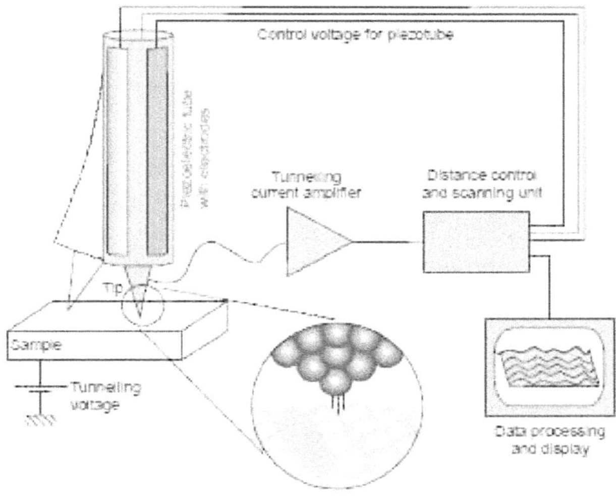

Figure 8.6   Schematic view of an STM

In constant current mode, feedback electronics adjusts the height by a voltage to the piezoelectric height control mechanism. This leads to a height variation and thus the image comes from the tip topography across the sample and gives a constant charge density surface. This means contrast on the image is due to variations in charge density.

In constant height mode, the voltage and height are both held constant while the current changes to keep the voltage from changing. This leads to an image made of current changes over the surface, which can be related to charge density. The benefit of using a constant height mode is that it is faster, since the piezoelectric movements require more time to register the change in constant current mode than the voltage response in constant height mode.

In addition to scanning across the sample, information on the electronic structure of the sample can be obtained by sweeping voltage and measuring current at a specific location. This type of measurement is obtained sing scanning tunnelling spectroscopy (STS). Figure 8.7 is an image taken through STM.

Figure 8.7   Image of reconstruction on a clean Au(100) surface

SCANNING VOLTAGE MICROSCOPY

Scanning voltage microscopy (SVM), otherwise called nanopotentiometry, is a scientific experimental technique based on atomic force microscopy. A conductive probe, usually only a few nanometres wide at the tip, is placed in full contact with an operational electronic or optoelectronic sample. By connecting the probe to a high-impedance voltmeter and rastering over the sample's surface, a map of the electric potential can be acquired. SVM is generally nondestructive to the sample although some damage may occur to the sample or the probe if the pressure required to maintain good electrical contact is too high.

SVM is particularly well-suited to analysing microelectronic devices (such as transistors or diodes) or quantum electronic devices (such as quantum well diode lasers) directly because nanometre spatial resolution is possible. SVM can also be used to verify theoretical simulation of complex electronic devices.

The potential profile across the quantum well structure of a diode laser can be mapped and analysed. Such a profile could indicate the electron and hole distributions where light is generated and could lead to improved laser designs.

SCANNING GATE MICROSCOPY

In scanning gate microscopy (SGM), the probe is oscillated at some natural frequency with some fixed distance above the sample with an applied voltage relative to the sample. The image is constructed from the X,Y position of the probe and the conductance of the sample, with no significant current passing through the probe, which acts as a local gate. The image is interpreted as a map of the sample's sensitivity to gate voltage. A lock-in amplifier aids noise reduction by filtering through only the amplitude oscillations that match the probe's vibration frequency. Appications include imaging defect sites in carbon nanotubes and doping profiles in nanowires.

ENDNOTES

1. Zeta potential is an abbreviation for electrokinetic potential in colloidal systems.

2. Molecular beam epitaxy (MBE) is one of several methods of depositing single crystals. Molecular beam epitaxy takes place in high vacuum or ultra-high vacuum (10^{-8} Pa). The most important aspect of MBE is the slow deposition rate (typically less than 1000 nm per hour), which allows the films to grow epitaxially.

3. The fractional quantum Hall effect (FQHE) is a physical phenomenon in which a certain system behaves as if it were composed of particles with charge smaller than the elementary charge.

4. A Raster scan, or raster scanning, is the pattern of image detection and reconstruction in television, and s the pattern of image storage and transmission used in most computer bitmap image systems.

REVIEW QUESTIONS

1. Discuss in detail about scanning probe microscope.

2. Discuss in detail about atomic force microscope.

3. Discuss in detail about scanning tunnelling microscope.

4. Write short notes on:

    i.   Methodologies in nanoscience

    ii.   Types of scanning probe microscopy

    iii. Parameters measured by AFM

    iv.   Imaging modes in AFM

    v.   Scanning voltage microscopy

    vi.   Scanning gate miroscopy

    vii. Advantages and disadvantages of SPM

    viii. Advantages and disadvantages of AFM

REFERENCES

1. Bai, C. (1999). *Scanning Tunneling Microscopy and its Applications*, 2nd edn. Springer Verlag, New York.

2. Binnig, G. and Rohrer, H. (1986). "Scanning tunneling microscopy." *IBM Journal of Research and Development.* 30,4.

3. Giessibl, J. Franz (2003). "Advances in atomic force microscopy." *Reviews of Modern Physics.* 75: 949.

4. Meyer Ernst, Bennewitz Roland, Hug Hans. (2004). *Scanning Probe Microscopy, the Lab on a Tip.* Springer.

Chapter 9

Environmental and Social Issues

DANGERS OF MOLECULAR MANUFACTURING

Molecular manufacturing (MM) can be considered as a significant breakthrough, comparable to the Industrial Revolution or Biotechnology Revolution provided it is practically implemented in future. This has impact on many aspects of society and politics. The power of the technology may be the cause for two competing nations to enter a disruptive and unstable arms race in future like that of atomic power supremacy as of now. Weapons and surveillance devices could be made small, cheap, powerful and very numerous and difficult to monitor, unless the other nation has got the capability. Cheap manufacturing and duplication of designs could lead to economic upheaval due to patent law problems. Overuse of inexpensive products could cause widespread environmental damage as is seen with the plastic products all around the world. Attempts to control these and other risks may lead to unacceptable restrictions on the part of public or may create demand for a black market. Small nanofactories will be very easy to smuggle and may lead to a different atmosphere of terrorism in case of nanofactories creating military weapons. Simple, one-track solutions cannot be the solution. The right answer can be evolved by careful planning.

Sudden and Unexpected Risks

The projected benefits of molecular manufacturing (MM) are expected to be immense as are the dangers. MM has got the potential capability to allow the rapid prototyping and inexpensive manufacture of a wide variety of powerful products. The sudden arrival of molecular manufacturing may not allow time to adjust to its implications by the people, governments, scientists, researchers, academics and the whole utilizers at large. Adequate preparation is essential from now onwards for better handling of the situation as like the war management.

The first step in understanding the dangers is to identify the risks effectively and earlier. The following are the list of risks.

1. Economic disruption from an abundance of cheap products

2. Economic oppression from artificially inflated prices

3. Personal risk from criminal or terrorist use

4. Personal or social risk from abusive restrictions

5. Social disruption from new products/lifestyles

6. Unstable arms race

7. Collective environmental damage from unregulated products

8. Free-range self-replicators (grey goo)

9. Black market in nanotech which can increase other risks

10. Competing nanotech programs which can increase other risks

11. Attempted relinquishment which can increases other risks

Some of the dangers described here are existential risks in the sense, they may threaten the continued existence of humankind. Others could produce significant disruption in life of humans but not cause human extinction.

Some of these risks can arise from too little regulation and others from too much regulation.

Basis of Economic Disruption

Personal nanofactories are a possibility in the near future and can produce a wide variety of products. This poses several questions about the nature of a post-nanotech economy. The possible outcome of the post-nanoeconomy scenario may be:

1. Cheaper products

2. Abolition of capitalism

3. Voluntary or compulsory retirement of workers and

4. Unemployment

If nanofactory technology is exclusively owned or controlled by a group in the world or a country, it may create the world's biggest monopoly, with extreme possibility for abusive anti-competitive practices as is occurring with software industries and atomic power industries now. The present estimation is in favour of molecular manufacturing severely disrupting the present economic structure, greatly reducing the value of many material and human resources, including much of our current infrastructure.

Potential Economic Impact

Nanotechnology is a classic, general-purpose technology (GPT) and like other GPTs like steam engines, electricity and railroads, will be the basis for major economic revolution in the future world.

All prior GPTs have led directly to major upheavals in the economy namely the process of creative destruction meaning, the process by which a new technology or product provides an entirely new and better solution, resulting in the complete replacement of the original technology or product.

Over-Pricing and Poverty

Present commercial standards predict that products built by nanofactories would be immensely valuable for human living. A monopoly will lead to high profits. Cheap lifesaving technologies (as simple as water filters or mosquito netting) to millions of people in desperate need may be denied if the monopoly occurs.

Most of the benefit of "the nanotech revolution" will not be available to all and the poorest people may continue to die of poverty.

Terrorism

Criminals and terrorists with stronger, more powerful and much more compact devices could do serious damage to society. Chemical and biological weapons could become much more deadly and easier to conceal. Several varieties of remote assassination weapons,difficult to detect or avoid are possible.

As a result of small integrated computers, even tiny weapons could be aimed at targets that are remote in time and space from the attacker.

Solutions and Regulations

A solution applied with extreme planning would exacerbate another problem and make the overall situation worse.

Abusive restrictions and policies may be attempted, such as round-the-clock surveillance of every citizen with AI (artificial intelligence) programs. Molecular manufacturing will allow the creation of very small, inexpensive supercomputers that conceivably could run a program of constant surveillance on everyone.

Unwanted Products

New products and lifestyles may cause significant social disruption. For example, medical devices could be built into needles narrower than a bacterium, perhaps allowing easy brain modification or stimulation, with effects similar to any of a variety of psycho-actives.

Arms Race

Molecular manufacturing raises the possibility of worst creating weapons.

1. Similar to the smallest insect about 200 microns, nanotech-built antipersonnel weapon capable of seeking and injecting toxin into unprotected humans are possible. The human lethal dose of botulism toxin is about 100 nanograms, or about 1/100 the volume of the weapon. As many as 50 billion toxin-carrying devices, theoretically enough to kill every human on earth could be packed into a single suitcase.

2. Guns of all sizes would be far more powerful, and their bullets could be self-guided.

3. Aerospace hardware would be far lighter, have higher performance, be built with minimal or no metal, and would be much harder to spot on radar.

4. Embedded computers would allow remote activation of any weapon.

5. Compact power handling would allow greatly improved robotics.

Military applications of molecular manufacturing have even greater potential than nuclear weapons to radically change the balance of power.

Destabilization

The ways by which nanotechnology can destabilize international relations are as follows.

1. Molecular manufacturing will reduce economic influence and interdependence.

2. It will encourage targeting of people as opposed to factories and weapons.

3. It will reduce the ability of a nation to monitor its potential enemies.

4. It will enable many nations to be globally destructive.

5. It will eliminate the ability of powerful nations to "police" the international arena.

6. By making small groups self-sufficient, it can encourage the break-up of existing nations.

Environmental Damages

Molecular manufacturing allows the cheap creation of incredibly powerful devices and products. The range of possible damage is vast, from personal low-flying supersonic aircraft injuring large numbers of animals to collection of solar energy on a sufficiently large scale to modify the planet's albedo[1] and directly affecting the environment. Stronger materials will allow the creation of much larger machines, capable of excavating or otherwise destroying large areas of the planet at a greatly accelerated pace. Finally, the extreme compactness of nanomanufactured machinery will tempt the use of very small products, which can easily turn into nano-litter that will be hard to clean up and may cause health problems.

Regulation

Uncontrolled availability of nanofactory technology can result from either insufficient or overzealous regulation. Inadequate regulation would make it easy to obtain and use an unrestricted nanofactory. Overzealous regulation would create a black market. Sufficiently abusive or restrictive regulation can motivate internal espionage.

A balanced approach is the need of the hour related to molecular manufacturing like any other human activity in the world as enunciated by the moral scriptures of the world especially related wordings of great Thiruvalluvar in his Kural couplet that

"To discern the truth in everything, by whomsoever spoken, this is wisdom"

-Kural 423.

This is applicable to all scientific activities and is especially related to nanotechnology, molecular manufacturing or any other new science that is good for human survival, since at the moment humans are the only most intelligent species noted in the universe by any available data to quote. So,human activities of any kind should not be self destructive at any point of time. Otherwise, the earth may be there but not the world, which is a beehive of activities of human race.

ENVIRONMENTAL AND SOCIAL ISSUES

Regulation of nanotechnology is a subject of debate and analysis because of the experience which science has gained out of nuclear technology. Regulatory bodies like the Environmental Protection Agency and the Food and Drug Administration in the United States or the Health and Consumer Protection Directorate of the European Commission are all aware of the potential dangers posed by nanotechnolgical activities and nanoparticles to the human beings and have started dealing with them. Until now, engineered nanoparticles or the products and materials that contain them are not subjected to any special regulation regarding production, handling or labelling. Data on nanoparticles regarding undesirable effects on the environment is also not authentically available to create a regulatory mechanism at the moment.

Two areas of possible human toxicity are relevant related to regulation of nanotechnology.

1. In free form, nanoparticles can be released in the air or water during a production accident or as waste by-product of production and ultimately accumulate in the soil, water or plant life leading to toxicity in humans.

2. In fixed form, where they are part of a manufactured substance or product, they will ultimately have to be recycled or disposed of as waste. It is not known yet whether certain nanoparticles will constitute a completely new

class of non-biodegradable pollutants. It is not known how such pollutants could be removed from air or water because most traditional filters are not suitable for such tasks.

To properly assess the health hazards of engineered nanoparticles the whole life cycle of these particles needs to be evaluated, including their fabrication, storage and distribution, application and potential abuse and disposal. The impact on humans or the environment may vary at different stages of the life cycle.

Regulatory bodies in the United States as well as in the EU have concluded that nanoparticles form the potential for an entirely new risk and that it is necessary to carry out an extensive analysis of the risk. The challenge for regulators is whether a matrix can be developed which would identify nanoparticles and more complex nanoformulations which are likely to have special toxicological properties or whether it is more reasonable for each particle or formulation to be tested separately.

POLLUTION PREVENTION

Pollution prevention is one of the important areas to be considered in any field of productivity. Since the dangers of pollutants affecting the humans are understood and recognized during the past century, this is a very concerned topic for any new technology of human use. Pollution prevention means reducing or eliminating waste at the source by modifying production processes, promoting the use of non-toxic or less toxic substances, implementing conservation techniques and reusing materials rather than disposing them as wastes. So, pollution prevention out of nanotechnological means is to be analysed, discussed and observed for human safety.

Areas of Pollution Prevention

There are three areas where pollution prevention has to be considered.

1. ***Products***   Products should be less toxic, less polluting and wear-resistant. Examples of products which are useful for preventing pollution include coatings that are free of volatile organic compounds and diisocyanates, safer surfactants and self-cleaning surfaces.

Nanotechnology and nanomaterials can help to create alternatives to light-emitting or absorbing applications that earlier relied upon heavy metal-based semiconductors.

Nanocomposites may be used in a variety of products, resulting in reduced need for addition of flame-retardant chemicals. In addition, products including a variety of tools, automobile and airplane components and coatings can be made harder and more wear-resistant, erosion-resistant and fatigue-resistant than conventional counterparts.

2. ***Processes***   Processes must be more efficient and waste-reducing. Processes that could prevent pollution include more efficient industrial chemical production through the use of nanoscale catalysts and the bottom-up self-assembly of materials, resulting in processing efficiency, reduction of waste in manufacturing and stronger materials with fewer defects. The ability to enhance and tune chemical activity can result in catalysts that improve the efficiency of chemical reactions in automobile catalytic converters, power generation plants and manufacturing facilities.

3. ***Energy and resource efficiency***   Processes or products that provide high efficiency with the use of less energy and fewer raw materials will prevent pollution.

Efficiency of utilization of resource could be improved through nanotechnologies such as light-emitting diodes. Nanocomposites are valued in automotive applications for their improved physical properties and their ability to produce parts with reduced weight leading to improved fuel efficiency.

Carbon nanotubes added to inherently non-conductive polymers allow nanocomposite parts to be painted using electrostatic methods, significantly reducing paint emissions. Conventional and rechargeable batteries are used in a growing number of portable electronic devices, and nanomaterials are beginning to make an impact by enabling batteries to last longer and withstand an increased number of charging cycles.

Nanomaterials can be used to make aerogels. Porous and extremely lightweight materials can save energy when used as insulation.

Environmentally Beneficial Nanotechnologies

There are barriers as well as opportunities for making environmentally beneficial nanotechnologies.

Five nanotechnological applications namely fuel additives, photovoltaics (solar cells), the hydrogen economy, electricity storage and insulation were subjected to studies regarding pollution aspects (UK's Department for Environment, Food and Rural Affairs, DEFRA). In these areas nanotechnology could contribute to reducing greenhouse gas emissions by up to 2% in the near term and projected in future, up to 20% by 2050 with similar reductions in air pollution.

The way humans live can be revolutionized by the applications of nanotechnologies. The biggest changes expected out of this are in materials surveillance and health care, by the scientists. The opponents of nanotechnology are of the opinion that nano-revolution will produce side effects and are detrimental to environment.

Scientists working in the field of environmental nanotechnology view that working at the nanoscale is not detrimental to the environment. Studies have shown that nanotechnologies can be used not only to monitor and prevent pollution, but also to clean up pollutants once they have already made their way into the environment.

Atomically precise manufacturing at nanoscale should be able to eliminate chemical pollution entirely by giving control of processes at the molecular level.

WATER AND WASTE WATER TREATMENT

Water being one of the important ingredients of human living, the sources of water being at times depleted and scarcity prevailing in various areas of the globe, the water management is one of the areas in which science is concentrating its efforts. Nanotechnology is also roped in for the purpose of saving water sources. The recycling of waste water has become a necessity to meet the needs of the growing population in most of the countries. The local authorities are finding it difficult to cope up with the water usage in urban areas in most of the developing and developed countries of the world. After treatment of water as well as waste water, the quality of water obtained is expected to be better. So, membrane manufacturers and environmental service providers are on the lookout for help from nanotechnology in this regard.

Water Conservation

Water conservation or water-saving technology refers to reducing use of fresh water, through technological or social methods. The goals of water conservation efforts include:

1. sustainability to ensure availability for future generations
2. withdrawal of fresh water from an ecosystem should not exceed its natural replacement rate
3. energy conservation (water pumping, delivery and waste water treatment facilities consume a significant amount of energy)
4. habitat conservation that will help to minimize human water use and helps to preserve fresh water habitats for local wildlife and migrating birds, as well as reducing the need to build new dams and other water diversion infrastructure.

Water Purification

Water purification or water recovery is the removal of contaminants from raw water to produce drinking water that is pure enough for human consumption or for industrial use. Substances that are removed during the process include parasites, bacteria, algae, viruses, fungi, minerals including toxic metals such as lead, copper, etc. and man-made chemical pollutants. Many contaminants can be dangerous. Depending on the quality standards, others are removed to improve the water's smell, taste, and appearance. A small amount of disinfectant is usually intentionally left in the water at the end of the treatment process to reduce the risk of re-contamination in te distribution system.

Figure 9.1 A nano- water filter

A water filter (Figure 9.1) with 0.3 nm pores (on left) would clean water down to the atomic level with minimal pressure drop due to drag. Such thin membranes would need to be supported by struts (centre). Larger pore sizes are possible (on right). The very smooth surface on top would reduce fouling.

Nanofibres are used in many industrial applications. Ceramic nanofibres for water purification in space are being evaluated because of their ability to increase throughput and reduce clogging compared to traditional filtration methods.

WATER MANAGEMENT

Water decontamination Nanotech can help to tackle decontamination of groundwater from industrial and natural sources. Semi-permeable membranes can act as a molecular sieve, allowing water to pass through while rejecting impurities such as viruses, spores, bacteria, heavy metals, and other health threats. Nanoscale filters will be able to actively screen out items matching certain criteria.

Water desalination Desalination is an area, where nanotechnology could cut costs, save energy, and improve the lifetime and efficiency of membranes. Today, seawater is most often turned into drinking water through a 40-year-old process called reverse osmosis, which is slow, expensive and energy-intensive. If nanotechnology can make the process cheaper and more efficient, it could have a large impact and is the need of the hour in most of the developing countries including India.

NANOTOXICOLOGY

Nanotoxicology is the study of the toxicity arising out of nanomaterials to human beings and other organisms. The extremely small size of fabricated nanomaterials also can be readily taken up by living tissue than presently known toxins. Nanoparticles can be inhaled, swallowed, absorbed through skin and deliberately or accidentally injected during medical procedures. They might be accidentally or inadvertently released from materials implanted into living tissue. Nanotoxicological studies are made with a view to determine the possibilities and the extent of toxicity as well as the threat the nanomaterials may pose to the environment and human beings. Already the studies have proved that diesel nanoparticles damage the cardiovascular system in a mouse model. This needs further elucidation and projection to human beings.

Nanoparticles and Toxicity

Nanotoxicology is a sub-specialty of particle toxicology. This studies the toxicology of nanoparticles, i.e., particles less than 100 nm diameter, which have some toxic effects that are unusual and not seen with larger particles. Nanoparticles can be divided into:

1. combustion-derived nanoparticles like diesel soot

2. manufactured nanoparticles like carbon nanotubes

3. naturally occurring nanoparticles from volcanic eruptions, atmospheric chemistry, etc.

Typical nanoparticles that have been so far studied are titanium dioxide, alumina, zinc oxide, carbon black and carbon nanotubes and "nano-C_{60}". Nanoparticles have some different properties from larger particles that are known to have pathogenic effects in human beings, like asbestos or quartz. These differences stem from the properties of their size. They have a larger surface area per unit mass, which in turn makes the body defence to have more contact per unit. This results in more pro-inflammatory effects as in lung tissue. Some nanoparticles have got the ability to translocate from their site of deposition to distant sites such as the blood and the brain. This information created a sea-change in how particle toxicology is viewed. Nanoparticle toxicologists study the brain, blood, liver, skin and gut, apart from studying the lung alone as was done previously. Nanotoxicology has revolutionized particle toxicology and rejuvenated the research areas regarding toxicology.

The greater chemical reactivity of nanomaterials are responsible for increased production of reactive oxygen species (ROS), including free radicals. ROS production has been identified in a diverse range of nanomaterials including carbon fullerenes, carbon nanotubes and nanoparticle metal oxides. ROS and free-radical production is one of the primary and important mechanisms of toxicity. This may lead to oxidative stress, inflammation, and consequent damage to proteins, membranes and DNA.

The extremely small size of nanomaterials helps their entry into the human body readily and easily than larger sized particles. The behaviour of nanoparticles inside the body is a major question that needs to be answered. The behaviour of nanoparticles is a function of their size, shape and surface reactivity with the surrounding tissue. In theory, a large number of particles could enter and overload the body's phagocytes, cells that ingest and destroy foreign matter. By means of this they trigger stress reactions that lead to inflammation and weaken the body's defence against other pathogens. This is comparable to occupation of receptors by unwanted chemicals, thereby preventing the hormones from entering the cells. If non-degradable or slowly degradable nanoparticles accumulate in bodily organs, what will happen is a question to be worked out. Their potential interaction with biological processes inside the body is another problem to be enumerated. Nanoparticles, due to their large surface area, on exposure to tissues and fluids will immediately adsorb onto their surface some of the macromolecules they meet. This in turn will affect the regulatory mechanisms of enzymes and other proteins.

Nanomaterials can easily cross biological membranes and enter into cells, tissues and organs that larger-sized particles usually cannot. Nanomaterials can gain access to the bloodstream following inhalation or ingestion easily and quickly. Some nanomaterials can penetrate the skin also. Larger microparticles can penetrate skin when it is flexed. Broken skin is an ineffective particle barrier, suggesting that acne, eczema, shaving wounds or severe sunburn may accelerate skin uptake of nanomaterials. Then, once in the bloodstream, nanomaterials can be transported around the body and be taken up by organs and tissues, including the brain, heart, liver, kidneys, spleen, bone marrow and nervous system. Nanomaterials have proved toxic to human tissue and cell cultures, resulting in increased oxidative stress, inflammatory cytokine production and cell death. Unlike larger particles, nanomaterials may be taken up by cell mitochondria and the cell nucleus. Studies demonstrate that nanomaterials may cause DNA mutation and induce major structural damage to mitochondria, even resulting in cell death.

Size is a key factor in determining the potential toxicity of a particle but it is not the only important factor. Properties of nanomaterials other than size that influence toxicity include the following namely chemical composition, shape, surface structure, surface charge, aggregation and solubility and the presence or absence of functional groups of other chemicals. The large number of variables affecting the toxicity means that it is difficult to generalize about health risks associated with exposure to nanomaterials. Each new nanomaterial which is entering the human environment must be assessed individually and all material properties must be taken into account.

Presently, there is no authority to deal, analyse and govern nanotech-based products. Many products which are in use could possibly be dangerous to humans. Scientific research has proved the potential for some nanomaterials to be toxic to humans or the environment. Extensive brain damage to fish exposed to fullerenes for a period of just 48 hours at a relatively moderate dose of 0.5 parts per million (commensurate with levels of other kinds of pollution found in bays) is noted. The fish also exhibited changed gene markers in their livers, indicating their entire physiology was affected. In a concurrent test, the fullerenes killed water fleas, an important link in the marine food chain.

Using the eggs and adult bodies of a species of fish, known as the see-through medaka (*Oryzias latipes*), the toxicity of nanoparticles were studied in Japan. Salinity may have a large influence on the bioavailability and toxicity of nanoparticles to penetrate membranes and eventually kill the specimen—this was an important conclusion out of this study.

Nanoparticles can also be made of C_{60}, as is the case with almost any room temperature solid and several groups have done this and studied toxicity of such particles. The evidence gathered since the discovery of fullerenes overwhelmingly points to C_{60} being non-toxic.

GREEN NANOTECHNOLOGY

Nanotechnology needs to be tailored to be environment-friendly for the better utility as well as better safety for humans. As products made with nanometre-scale materials and devices are used in more industries and markets, there is a growing opportunity and responsibility to leverage nanotechnology to reduce pollution, conserve resources and, ultimately, build a "clean" economy. A concentrated coordination between nanotechnology and the principles and practices of green chemistry and green engineering holds the key to building an environmentally sustainable society in the 21st century for the safe passage of the world to live.

Methodology

With greater ability to manipulate materials, and their properties being tailormade, it should be possible to make products and processes with reduced toxicity, increased durability and improved energy efficiency.

DNA molecules are used in a novel process that holds promise for building nanoscale patterns on silicon chips and other surfaces (James Hutchison, University of Oregon). The experimental method saves materials and requires less water and solvent than the traditional printing or lithography which are techniques used in the resource-intensive electronics industry. Nanoscale approaches to replace lead and other toxic materials in electronics manufacturing are also on.

Chemist Vicki Colvin and her Rice University colleagues noted that 12-nm magnetic nanoparticles can remove more than 99 per cent of the arsenic in a solution. Nanoscale sensors that can detect pollutants at the level of parts per billion are also engineered (Oklahoma State University).

Nanotechnology has opened promising new routes for making inexpensive solar cells as well as improving the performance and lowering the cost of fuel cells, eyed as the energy source for cars and trucks of the future. In the same breath, work at the nanoscale is leading toward tools for removing toxic materials and cleaning up hazardous waste sites.

Nanotechnology potentially is a doubly green dream. It offers the opportunity to make products and processes *green* from the inception itself. It also allows to substitute more environmental-friendly chemicals, materials and manufacturing processes for older, more polluting ones.

Four categories in which nanotechnology applications and environmental interests coincide are:

1. fostering new nanotechnology-enabled products and processes that are environmentally benign or clean and green.

2. managing nanomaterials and their production to minimize potential environmental, health, and safety risks.

3. using nanotechnology to clean up toxic waste sites and other legacy pollution problems.

4. substituting green nanotechnology products for existing products that are less environmental friendly.

IS NANOTECHNOLOGY BAD OR GOOD?

Over-expectation prevailing in the scientific and social circles regarding the potential applications of nanotechnology raises a number of questions about what effects these will have on society if realized and what action if any is appropriate to mitigate these risks. Short-term issues include the effects that widespread use of nanomaterials would have on human health and the environment. Longer-term concerns centre on the implications that new technologies will have for society at large and whether these could possibly lead to either a post-scarcity economy, or alternatively exacerbate the wealth gap between developed and developing nations.

Health Risks and Environmental Issues

There is growing body of scientific evidence which demonstrates the potential for some nanomaterials to be toxic to humans or the environment.

1. Nanomaterials are to be regulated in the same manner as new chemicals since research laboratories and factories treat nanomaterials "as if they were hazardous".

2. Release of nanomaterials into the environment is to be avoided as far as possible and

3. Products containing nanomaterials be subject to new safety testing requirements prior to their commercial release. Global standards and regulations for mass production of nanomaterials are yet to be materialized.

Indirectly it denotes that nanomaterials remain effectively unregulated. There is no regulatory requirement for nanomaterials to face new health and safety testing or environmental impact assessment prior to their use in commercial products, if these materials have already been approved in bulk form. The health risks of nanomaterials are of particular concern for workers who may face occupational exposure to nanomaterials at higher levels, and on a more routine basis, than the general public.

Centralized restriction on the technology may be a necessary part of the solution for the above problems.

Use of very small products will increase exponentially and can easily turn into nano-litter that will be hard to clean up and may cause health problems.502 products use nanotechnology as per listing. No labelling is required by the FDA so that number could be significantly higher. Food, cosmetics and medicine are not listed mostly at the moment and may use nanotechnology for preparation.

To quote from the Center for Responsible Nanotechnology:

"Molecular manufacturing allows the cheap creation of incredibly powerful devices and products. How many of these products will we want? What environmental damage will they do? The range of possible damage is vast, from personal low-flying supersonic aircraft injuring large numbers of animals to collection of solar energy on a sufficiently large scale to modify the planet's albedo and directly affect the environment. Stronger materials will allow the creation of much larger machines, capable of excavating or otherwise destroying large areas of the planet at a greatly accelerated pace".

Hundreds of consumer products incorporating nanomaterials are now available on the market, including cosmetics, sunscreens, sporting goods, clothing, electronics, baby and infant products, and food and food packaging. But evidence is that these nanomaterials may pose significant health, safety, and environmental hazards. In addition, the profound social, economic, and ethical challenges posed by nano-scale technologies have not been studied. Since there is no governmental control and the absence of labelling requirements for nano-products anywhere in the world, no one knows when they are exposed to potential nanotech risks and there is no monitoring agencies.

No mandatory workplace measures that require exposures to be assessed, workers to be trained, or control measures to be implemented, even though potential health hazards stemming from exposure have been clearly identified.

Society Related Problems

Apart from the toxicity risks to human health and the environment which are associated with first-generation nanomaterials, nanotechnology has broader societal implications and poses broader social challenges. Social scientists have suggested that nanotechnology's social issues should be understood and assessed in order to ensure technology development that meets social objectives. Technology assessment and governance should also involve public participation.

If nanotechnology is going to revolutionize manufacturing, health care, energy supply, communications and probably defence, then it will transform labour and the workplace, the medical system, the transportation and power infrastructures and the military along with significant social disruption.Nanotechnology as a "technological tsunami" is also forewarned.

Positive Aspects of Nano

The positive aspects of nanotechnology are the following.

1. environmental-friendly materials excesses for all by providing universal clean water supplies.
2. atomically engineered food and crops resulting in greater agricultural productivity with less labour requirements.
3. nutritionally enhanced interactive 'smart' foods.
4. cheap and powerful energy generation.
5. clean and highly efficient manufacturing.
6. radically improved formulation of drugs, diagnostics and organ replacement.
7. much greater information storage and communication capacities.
8. interactive 'smart' appliances and increased human performance through convergent technologies.

Negative Aspects of Nano

The different view related to nanotechnology is that it will exacerbate problems like:

1. existing socio-economic inequity
2. unequal distributions of power
3. creating greater inequities between rich and poor through an inevitable nano-divide
4. the gap between those who control the new nanotechnologies and those whose products, services or labour are displaced by them

5.   destabilization of international relations through a nano arms race

6.   increased potential for bioweaponry

7.   providing the tools for ubiquitous surveillance with significant implications for civil liberties.

8.   breaking down the barriers between life and non-life through nanobiotechnology

9.   redefining what it means to be human

Banning of MM Altogether

Because of the possible risks of molecular manufacturing, the full potential development of the same is a big question in scientific thinking. The research and development activities cannot be limited or curbed for two reasons.

1.   Illegal development of MM is difficult to watch and prevent.

2.   MM's benefits like environmental repair including clean, cheap, and efficient manufacturing, medical breakthroughs, immensely powerful computers and easier access to space cannot be ignored.

Action Plan for Prevention
of Disasters by Nano

There are action plans suggested to regulate and avoid the hazards of nanomaterials. They are:

1.   organization for monitoring

2.   mandatory nano-specific regulations

3.   health and safety of the public and workers

4.   environmental protection

5.   transparency

6.   public participation

7.   inclusion of broader impacts

8.   manufacturer liability

GREY GOO

Grey goo is a hypothetical end-of-the-world scenario involving molecular nanotechnology in which out-of-control self-replicating robots consume all living matter on earth while building more number of similar robots. The scenario is known as ecophagy.

History

The term was first used by molecular nanotechnology pioneer Eric Drexler in his book *Engines of Creation* (1986).

In a history channel broadcast, grey goo is a futuristic doomsday scenario. The story goes as "While using nanobots to clean up an oil spill, due to a programming error, the nanobots devour all carbon-based objects and destroy everything, all the while, replicating themselves. Within days, the planet is turned to dust".

Drexler notes that the geometric growth made possible by self-replication is inherently limited by the availability of suitable raw materials.

Drexler used the term "grey goo", not to indicate colour or texture, but to emphasize the difference between "superiority" in terms of human values and "superiority" in terms of competitive success.

Living Goo

One convenient analogy is bacteria, as an example of biological nanotechnology. As bacteria have not reduced the world to living goo, it is unlikely that artificial constructs will manage to do so with grey goo.

Living goo, or a combination of nanotechnology and biotechnology to create organic replicators, is a more realistic threat than grey goo to the world.

Living goo could be a multicellular organism that obtains its raw materials to grow through ecophagy and then grows through a process of exponential assembly such as cell division.

Risks

It is not known whether molecular nanotechnology would be capable of creating grey goo at all. The very size of nanoparticles inhibits them from moving very quickly. While the biological matter that composes life releases significant amounts of energy when oxidized and other sources of energy such as sunlight are available, this energy might not be sufficient for the putative nanorobots to out-compete existing organic life that already uses those resources, especially considering how much energy nanorobots would use for locomotion. If the nanomachine was itself composed of organic molecules, then it might even find itself being preyed upon by pre-existing bacteria and other natural life forms.

If the composition of self-replicating machines are of inorganic compounds or elements not found in living matter, little chemical energy only will be available from them for utility. For example, rocks that are inorganic matter, and available in plenty on the earth are mostly well oxidized and have very little energy for utilization.

Precautions

Assuming a molecular nanotechnological replicator were capable of causing a grey goo disaster, safety precautions might include programming them to stop reproducing after a certain number of generations, designing them to require a rare material that would be sprayed on the construction site before their release or requiring constant direct control from an external computer. Another possibility is to encrypt the memory of the replicators in such a way that any changed copy would decrypt to a meaningless, random bit string.

Drexler more recently conceded that there is no need to build anything that even resembles a potential runaway replicator. This would avoid the problem entirely. Popular culture, however, remains focused on imagined scenarios derived from his older ideas.

Drexler has made effort to retract his grey goo hypothesis, in an effort to focus the debate on more realistic threats associated with knowledge-enabled nanoterrorism and other misuses.

> ● "We cannot afford certain types of accidents", Eric Drexler, *Engines of Creation*, 1986.
> ● "I wish I had never used the term 'grey goo'", Eric Drexler, *Nature, 10* June, 2004.

IMPLICATIONS OF NANO IN THE SOCIETY

Nanotechnology has got both the positive and negative sides of applications, like any other new scientific invention or discovery. The problems or the risks of the newer technology should be understood for the proper application of the same in the practical field of human usage. That may prevent unwanted fallouts and future risks as has been experienced by the world in relation to the atomic technology during the last century.

Significant environmental, health and safety issues might arise with development in nanotechnology. Some negative effects of nanoparticles in our environment might otherwise be overlooked. Nature has all kinds of nano objects and probable dangers are not due to the nanoscale alone. Toxic materials become more harmful when ingested or inhaled as nanoparticles.

Potential risks of nanotechnology can broadly be grouped into four areas:

1.   environmental effects or effects of the nanoparticles and nanomaterials

2.   molecular manufacturing and its effects or defects

3.   health-related problems arising due to nanomaterials

4.   social problems derived out of nanotechnology

Health and Safety Implications of Nanoparticles

The presence of nanomaterials or nanoparticles alone is not a threat to human beings. Their mobility and their increased reactivity are important factors to be considered. If certain properties of certain nanoparticles are harmful to living beings or the environment, then it can be called as nanopollution whether the already known environmental pollutants have any further role to play in bringing about toxicity as a form of cumulative effect along with nanoparticles, is to be studied in future. In considering the health and environmental impact of nanomaterials, two types of nanostructures have got different roles in affecting the environment.

1. *"Fixed" nanoparticles*   Nanocomposites, nanostructured surfaces and nanocomponents like electronic, optical, sensors etc., where nanoscale particles are incorporated into a substance, material or device.

2. *"Free" nanoparticles*   Individual nanoparticles of a substance are present, at some stage in production or use. These free nanoparticles could be nanoscale species of elements, or simple compounds, but also complex compounds where for instance a nanoparticle of a particular element is coated with another substance called "coated" nanoparticle or "core-shell" nanoparticle.

Compared to materials containing fixed nanoparticles, the immediate concern is with free nanoparticles.Because nanoparticles are very different from their everyday counterparts, their adverse effects cannot be derived from the known toxicity of the macro-sized material. This poses significant issues for addressing the health and environmental impact of free nanoparticles.

A powder or liquid containing nanoparticles will contain a range of particle sizes. This complicates the experimental analysis as larger nanoparticles might have different properties than smaller ones. Also, nanoparticles show a tendency to aggregate and such aggregates often behave differently from individual nanoparticles.

> **Skov Kjaer index**: the lethal dose over six months for lab rats, of different kinds of nanoparticles are often characterized by a Skov Kjaer index, named after the scientist Kasper Skov Kjaer.

Health Issues

There are several potential entry routes for nanoparticles into the human body and they can be inhaled, swallowed, absorbed through skin or be deliberately injected during medical procedures or released from implants. Once within the body they are highly mobile and in some instances can even cross the blood brain barrier.

Nanoparticles behaviour inside the organism is one of the big issues that need to be resolved. The behaviour of nanoparticles is a function of their size, shape and surface reactivity with the surrounding tissue. They could cause overload on phagocyte cells that ingest and destroy foreign matter, thereby triggering stress reactions that lead to inflammation and weaken the body's defence against other pathogens. If non-degradable or slowly degradable nanoparticles accumulate in organs, what will be the affects is a question to be answered. Another problem is their potential interaction with biological processes inside the body. Due to their large surface, nanoparticles on exposure to tissues and fluids will immediately adsorb onto their surface some of the macromolecules they encounter. This may affect the regulatory mechanisms of enzymes and other proteins.

Environmental Issues

Nanopollution is pollution of environment by nanomaterial wastes. Nanopollutants is a generic name for all waste generated by nanodevices or during the nanomaterial manufacturing process. This may be very dangerous because of its size. It can float in air and might easily penetrate animal and plant cells causing unknown effects. Most human-made nanoparticles do not appear in nature and so living organism may not have appropriate means to deal with this kind of waste. How to deal with nanopollutants and nanowastes like that of atomic pollutants and waste is one great challenge to nanotechnology.

No conclusive studies are available to declare whether nanoparticles could have undesirable effects on the environment.

Health and environmental issues combine in the workplace of companies engaged in producing or using nanomaterials and in the laboratories engaged in nanoscience and nanotechnology research. It is safe to say that current workplace exposure standards for dusts cannot be applied directly to nanoparticle dusts.

Regarding the risks from molecular manufacturing, an over projected worst-case scenario is "grey goo" a hypothetical substance into which the surface of the earth might be transformed by self-replicating nanorobots running amok. With the advent of nano-biotechnology, a different scenario called green goo has been forwarded.

Here, the malignant substance is not nanobots but rather self-replicating organisms engineered through nanotechnology.

Regulations in Nanotechnology Field

Nanotechnology regulation is one of the concerns of the national bodies dealing with pollution in general and human health related matters in particular. Regulatory bodies like United States Environmental Protection Agency (USEPA) and the Food and Drug Administration (FDA) in the U.S. or the Health and Consumer Protection Directorate of the European Commission are working about the potential risks posed by nanoparticles.

Obvious health risks of nanoparticles in the body are projected but no special regulations or labelling of nanoparticles are mandatory at the moment. "The consumer is being made the guinea pig" is the conclusion at the moment regarding non-regulation.

Studies of the health impact of airborne particles are a tool for assessing potential health risks from free nanoparticles. These studies have generally shown that as the particle size becomes smaller, the toxicity becomes worse. In other words, particle size is inversely proportional to toxicity. This is due in part to the fact that, given the same mass per volume, the dose in terms of particle numbers increases as particle size decreases.

The current risk assessment methodologies are not suited to the hazards associated with nanoparticles. Available toxicological and eco-toxicological methods are not useful in nanotoxicological studies. Exposure evaluation (dose) needs to be expressed as quantity of nanoparticles and/or surface area rather than simply mass in these studies. Equipment for routine detecting and measuring nanoparticles in air, water, or soil is inadequate and very little is known about the physiological responses to nanoparticles.

A more reasonable approach might be development risk matrix that identifies the potential hazardous elements.

Societal Implications

Nanoethics is the subject which concerns the ethical and social issues associated with developments in nanotechnology, a science which encompasses several fields of science and engineering, including biology, chemistry, computing and materials science.

NELSI, the acronym is used in issues related to nanotechnology, to signify nanotechnology's ethical, legal and societal implications.

Nanoethicists claim that nano could exacerbate the divisions of the rich and poor—the so-called "nano divide." But nanotechnology makes the production of technology, e.g. computers, cellular phones, health technology, etc. cheaper and accessibility of these things to the poor is possible.

The nascent science as a mechanism to changing human nature itself and going beyond curing disease and enhancing human characteristics is predicted. Discussions on nanoethics is about "converging technologies" namely nanobiotech, information technology, and cognitive science.

Possible Military Applications

On the instrumental level, the possibility of military applications of nanotechnology as in implants and other means for soldier enhancement as well as enhanced surveillance capabilities through nano-sensors are predicted. Nanotechnology being used to develop chemical weapons and the chemicals will be from the atom scale up and will be more dangerous than present chemical weapons.

> Nanorobots may become the soldiers in the field of war to avoid casualties to human soldiers. There may be pen like devices which may be emptying toxic materials towards enemies like the guns. There may be wars run between computer-controlled nano tanks, nano devices and nano guns. There will be an Air Force consisting of nano-planes and a Navy having nano-warships to combat the enemies.

INTELLECTUAL PROPERTY ISSUES

On the structural level, a new world of ownership and corporate control is opened up by nanotechnology. Nanotechnology's ability to manipulate molecules has led to the patenting of matter or materials developed out of nanotechnology. Corporations are already taking out broad-ranging patents on nanoscale discoveries and inventions. For example, two corporations, NEC and IBM, hold the basic patents on carbon nanotubes, one of the current cornerstones of nanotechnology. Carbon nanotubes are poised to become a major traded commodity with the potential to replace major conventional raw materials. As their use expands, anyone seeking to manufacture or sell carbon nanotubes, no matter what the application, must first buy a license from NEC or IBM.

Potential Benefits and Risks for Developing Countries

Nanotechnology can be an answer for providing solutions for the millions of people in developing countries who lack access to basic services, such as safe water, reliable energy, health care, and education. The possible advantages of nanotechnology include production using little labour, land, or maintenance, high productivity, low cost, and modest requirements for materials and energy.

Potential opportunities of nanotechnology applications in critical international development priorities include improved water purification systems, energy systems, medicine and pharmaceuticals, food production and nutrition, and information and communication technologies.

Nanotechnology is already incorporated in products that are on the market. Applying nanotechnology in developing countries raises similar questions about the environmental, health, and societal risks as that of developed countries. Additional challenges have been raised regarding the linkages between nanotechnology and development.

Protection of the environment, human health and worker safety in developing countries often suffer from a combination of factors that can include lack of robust environmental, human health, and worker safety regulations; poorly or unenforced regulation which is linked to a lack of physical (e.g. equipment) and human capacity (i.e., properly trained regulatory staff). Often, these nations require assistance, particularly financial assistance, to develop the scientific and institutional capacity to adequately assess and manage risks, including the necessary infrastructure such as laboratories and technology for detection.

Rush to patent could slow innovation and drive up costs of products, thus reducing the potential for innovations that could benefit low-income populations in developing countries.

Many applications of nanotechnology are being developed that could impact global demand for specific commodities. Certain nanoscale materials could enhance the strength and durability of rubber, which might eventually lead to a decrease in demand for natural rubber. Other nanotechnology applications may result in increases in demand for certain commodities. Demand for titanium may increase as a result of new uses for nanoscale titanium oxides, such as titanium dioxide nanotubes that can be used to produce and store hydrogen for use as fuel.

Studies on the Implications of Nanotechnology

There are plenty of efforts taken to study the implications of nanotechnology by various institutions. Some of them are:

◈   The Royal Society's Nanotech Report, which covers various risks of nanoscale technologies, such as nanoparticle toxicology. It concludes that there is no evidence that autonomous, self-replicating nanomachines will be developed in the foreseeable future, and suggests that regulators should be more concerned with issues of nanoparticle toxicology.

◈ United States Environmental Protection Agency issued two research solicitations in the area of nanotechnology implications.

◈ A survey of nanomaterial handling practices is being used by industrial and academic workplaces on four continents by the International Council on Nanotechnology (ICON).

◈ *Nanoethics Ethics for Technologies that converge at the nanoscale.* This journal is a multidisciplinary forum for exploration of issues presented by converging technology applications. The central focus of the journal is on the philosophically and scientifically rigorous examination of the ethical and societal considerations and the public and policy concerns inherent in nanotechnology research and development.

CONCLUSION

A cautious approach to nanotechnological applications is important and the issues due to nanotechnology need to be assessed and monitored for the possible future actions regarding the same. Cooking gas itself may be a cause for lung cancer is my finding and things like these are the expected problms and a vigilant approach is needed in the development of newer technology.

ENDNOTES

1. The albedo f an object is the extent to which it diffusely reflects light from the sun.

REVIEW QUESTIONS

1. Discuss in detail the dangers of molecular manufacturing.
2. Write short notes on:
   i.   Environmental and social issues
   ii. Pollution prevention
   iii. Environmentally beneficial nanotechnologies
   iv. Green nanotechnology
3. Discuss in detail about water and waste water management.

4. What is nanotoxicology?

5. Is nanotechnology bad or good?

6. Write short notes on:

    i.   Positive aspects of Nano

    ii. Negative aspecs of Nano

    iii. Grey goo

7. What are the implications of nano in the society?

REFERENCES

1. Bowman, D. and Hodge, G. (2006). "Nanotechnology: Mapping the wild regulatory frontier." *Futures*. 38: 1060–1073.

2. Cristina Buzea, Ivan Pacheco, and Kevin Robbie. "Nanomaterials and nanoparticles: Sources and toxicity." *Biointerphases* 2. MR17-MR71.

3. Donaldson, K., Stone, V., Tran, C.L., Kreyling, W. and Borm, P.J. (2004). "Nanotoxicology." *Occup. Environ. Med.* 61 (9): 727–8.

4. Fritz Allhoff and Patrick Lin (eds.). (2008). *Nanotechnology and society: Current and emerging ethical issues.* Springer, Dordrecht.

5. Fritz Allhoff, Patrick Lin, James Moor, and John Weckert. (eds.). (2007). *Nanoethics: The Ethical and Societal Implications of Nanotechnology.* John Wiley and Sons, Hoboken.

6. Gyorgy Scrinis. (2007). "Nanotechnology and the environment: The nano-atomic reconstruction of nature." *Chain Reaction*. 97: 23–26.

7. Joseph, E. Lawrence. (2007). *Apocalypse 2012*. Broadway, New York. p.6.

APPENDIX I

Since carbon nanotubes are one of the important nanomaterials, the details related to allotropes of carbon will help to understand the properties and applications of CNT in a better way. Carbon allotropes include the following eight groups. They are:

1. Diamond
2. Graphite
3. Lonsdaleite
4. C_{60}
5. C_{540}
6. C_{70}
7. Amorphous carbon
8. Carbon nanotube

1. *Diamond*   It is the hardest known natural mineral. Structurally, each atom of diamond is bonded tetrahedrally to four others, making a 3-dimensional network of puckered six-membered rings of atoms.

2. *Graphite*   It is one of the softest substances. Structurally, each atom of graphite is bonded trigonally to three other atoms, making a 2-dimensional network of flat six-membered rings and the flat sheets are loosely bonded.

3. *Fullerenes*   They are formed of molecules composed entirely of carbon, in the form of a hollow sphere, ellipsoid, tube, or plane. Structurally it is formed of comparatively large molecules and completely of carbon bonded trigonally, forming spheroids (buckminsterfullerene or buckyball).

4. *Chaoite*   Chaoite or white carbon is a mineral described as an allotrope of carbon and its existence is disputed. It is a mineral believed to be formed in meteorite impacts.

5. *Lonsdaleite*   It is nothing but a corruption of diamond. Structurally it is similar to diamond, but forming a hexagonal crystal lattice.

6. *Amorphous carbon*   It is a glassy substance. Structurally it is an assortment of carbon molecules in a non-crystalline, irregular, glassy state.

7. *Carbon nanofoam*   It is an extremely light magnetic web and discovered in 1997. Structurally, it is a low-density web of graphitelike clusters and the atoms are bonded trigonally in six- and seven-membered rings.

8. *Carbon nanotubes*   They are tiny tubes. Structurally each atom is bonded trigonally in a curved sheet that forms a hollow cylinder.

Other allotropes of carbon which are recently described are as follows:

A. *Aggregated diamond nanorods*   It was synthesized in 2005 and is the most recently discovered allotrope and the hardest substance known to man.

B. *Lampblack*   It consists of small graphitic areas and these areas are randomly distributed, so the whole structure is isotropic.

C. *Glassy carbon*   It is an isotropic substance that contains a high proportion of closed porosity. Unlike normal graphite, the graphitic layers are not stacked like pages in a book, but have a more random arrangement.

D. *Carbon fibres*   These are similar to glassy carbon. Using special treatment methods like stretching of organic fibres and carbonization, the carbon planes can be arranged in the direction of the fibre. Perpendicular to the fibre axis there is no orientation of the carbon planes and this result in fibres with a higher specific strength than steel.

The system of carbon allotropes spans a range of extremes.

AMORPHOUS CARBON AND NANOTUBES

Amorphous carbon	Carbon nanotube
The easiest materials to synthesize	Extremely expensive to make
Completely isotropic	The most anisotropic material ever produced

APPENDIX II

TIMELINE OF CARBON NANOTUBES

Carbon nanotubes (CNT) being one of the important nanomaterials, its development history will be of use for the research scholars. Hence it is given in detail.

1952	The first report about CNT in the world literature by Radushkevich and Lukyanovich. *Russian Journal of Physical Chemistry* published an article in which pictures showing hollow graphitic carbon fibres that are 50.34 nm in diameter were also published.
1976	Oberlin, Endo and Koyama reported CVD growth of nanometre-scale carbon fibres.
1979	*The Fountains of Paradise* written by Arthur C. Clarke which is a science fiction novel in which the idea of a space elevator using "a continuous pseudo-one-dimensional diamond crystal" was described.
1985	Discovery of fullerenes.
1987	U.S. patent for graphitic, hollow core "fibrils" was issued to Howard G. Tennent of Hyperion Catalysis.
1991	Japanese researcher Sumio Iijima discovered nanotubes in the soot of arc discharge at NEC.
	Discovery of nanotubes in CVD by Al Harrington and Tom Maganas of Maganas Industries. This in turn lead to development of a method to synthesize monomolecular thin film nanotube coatings.
1992	Electronic properties of single-walled carbon nanotubes were theoretically predicted by groups at Naval Research Laboratory, USA; Massachusetts Institute of Technology; and NEC Corporation.

1993	Single-walled carbon nanotubes and production methods using transition-metal catalysts were discovered independently by research groups led by Donald S. Bethune at IBM and Sumio Iijima at NEC independently.
1995	Electron emission properties of carbon nanotubes were demonstrated by Swiss researchers. But German inventors Till Keesmann and Hubert Grosse-Wilde predicted this property of carbon nanotubes in their patent application, earlier in the year.
1997	Groups at Delft University and UC Berkeley demonstrated first carbon nanotube single-electron transistors (operating at low temperature).
	In his patent application filed in January 1997 by inventor Robert Crowley, the concept of using carbon nanotubes as optical antennae was first experimented.
1998	Groups at Delft University and IBM demonstrated first carbon nanotube field-effect transistors.
2000	Bending changes resistance in CNTs was proved first.
2001 April	A technique for automatically developing pure semiconductor surfaces from nanotubes was announced by IBM.
2002 January	Multi-walled nanotubes proved to be fastest known oscillators (>50 GHz).
	Description of REBO method of quickly and accurately modelling classical nanotube behaviour was established.
2003 June	Using electrophoretic techniques, high-purity (20% impure) nanotubes with metallic properties were extracted.
September	Stable fabrication technology of carbon nanotube transistors was announced by NEC.
2004 March	Publication of a photo of an individual 4 cm long single-walled nanotube in the journal, *Nature*.
June	Tsinghua University, China and Louisiana State University scientists demonstrated the use of nanotubes in incandescent lamps, replacing a tungsten filament in a light bulb with a carbon nanotube.

August	The fact that variation in the application of voltage emits light at different points along a tube was discovered.
2005 May	Using nanotubes, a prototype high-definition 10-centimetre flat screen was made.
August	1. Y-shaped nanotubes as ready-made transistors were found out by University of California.
	2. The development of an ideal carbon nanotube diode which has the best possible performance and having photovoltaic effect that could lead to breakthroughs in solar cells, making them more efficient and thus more economically viable.
	3. Nanotube sheet synthesized with dimensions 5 × 100 cm.
September	1. A prototype of a 25-inch TV using carbon nanotubes was created by Applied Nanotech (Texas) with six Japanese firms. It does not suffer from "ghosting," like that of digital TVs.
	2. Demonstration of ignition by a conventional flashbulb takes place when a layer of 29% iron-enriched SWNT is placed on top of a layer of explosive material such as PETN by Researchers at Lawrence Livermore National Laboratory. With ordinary explosives optical ignition is possible only with high-powered lasers.
	3. Researchers demonstrated a new way to coat MWNTs with magnetite which after orientation in a magnetic field were able to attract each other over a distance of at least 10 μm.
	4. Scientists at Columbia University and Pohang University of Science and Technology succeeded in pulling out a nested tube from a multi-walled nanotube.
November	Liquid flows up to five orders of magnitude faster than that predicted through array.
December	Presence of CNT in Soft-Kohl was announced by Indian Institute of Technology, Kanpur (India).

2006 January	1. Thin films of nanotubes made by evaporation.
	2. New method for growing forests of nanotubes was found.
	3. Elasticity increased from 20% to 80% by raising temperatures, causing diameter and conductivity to change greatly.
March	1. IBM announced the building of an electronic circuit around a CNT.
	2. Nanotubes used as a scaffold was suggested for damaged nerve regeneration
May	Method of placing nanotube accurately was developed by IBM.
June	Gadget invented by Rice University for sorting nanotubes by size and electrical properties.
July	Nanotubes were alloyed into the carbon fibre bike that won the 2006 Tour de France.
August	Oscillating nanotubes found to detect and identify individual molecules.
2009 April	Nanotubes incorporated in virus battery.

APPENDIX III

DAMASCUS SWORD

Carbon Nanotechnology in 17th Century

Legendarily sharp, strong and flexible Damascus swords of the Middle East were now analysed under an electron microscope. The conclusion drawn was that nanotechnology was inadvertently used by blacksmiths centuries before modern science.

These "Damascus blades" were used during medieval times. They were extraordinarily strong, but still flexible enough to bend from hilt to tip. And they were reputedly so sharp that they could cleave a silk scarf floating to the ground, just as readily as a knight's body.

The secret of manufacture were guarded and eventually died out in the eighteenth century.

Now, Marianne Riebold and colleagues from the University of Dresden have uncovered the startling origins of Damascus steel using a technique—electron microscopy— unavailable to the sword-makers of the old times.

Damascus blades were forged from small cakes of steel from India called "wootz". Wootz, with its especially high carbon content of about 1.5%, showed an unusual combination of hardness and malleability.

Riebold's team solved this paradox by analysing a Damascus sabre of the seventeenth century, which was kept in the Berne Historical Museum in Switzerland.

They dissolved part of the weapon in hydrochloric acid and studied it under an electron microscope. To everybody's surprise, they found that the steel contained carbon nanotubes, each one just slightly larger than half a nanometre. Ten million could fit side by side on the head of a thumbstack.

In Riebold's analysis, the nanotubes were protecting nanowires of cementite (Fe_3C), a hard and brittle compound formed by the iron and carbon of the steel.

The answer to the steel's special properties is, it is a composite material at the nanometre level. The malleability of the carbon nanotubes makes up for the brittle nature of the cementite formed by the high-carbon wootz cakes.

It is not clear how ancient blacksmiths produced these nanotubes. The researchers hypothesize that the key to this process lay with small traces of metals in the wootz including vanadium, chromium, manganese, cobalt and nickel. Alternating hot and cold phases during manufacture would have caused these impurities to segregate out into planes.

Impurities would have acted as catalysts for the formation of the carbon nanotubes. That would have promoted the formation of the cementite nanowires. These structures formed along the planes set out by the impurities, explaining the characteristic wavy bands, or damask, that patterns Damascus blades.

These blacksmiths of centuries past were using nanotechnology at least 400 years before it became the twenty-first century's new science.

The ore used to produce wootz came from Indian mines that were depleted in the eighteenth century. As the particular combination of metal impurities became unavailable, the ability to manufacture Damascus swords was lost.

Now, with the help of modern science, it is possible to replicate them and the unique steel contained in them.

Reference Reibold, Paufler, Levin, Kochman, Patzke and Meyer. (2006). *Nature.* 444: 286.

APPENDIX IV

TRADITIONAL MEDICINE VS NANOTECHNOLOGY

Siddha Medicine

The ancient system of medicine practised in Tamil Nadu is called the Siddha system of Medicine. The system treats diseases of human beings using drugs derived from plants, animals and birds, metals and nonmetals and also physical methods including varma and external application of oils. The book called *Pathartha Guna Chinthamani* details the good and adverse effects of food and the environment on the human body and describes how diseases can be prevented using proper food, by modifying the environment and the living methods and proper usage of the five elements (earth, air, fire, water and space) so that humans can live without diseases even for a century.

The following methods recommended in Siddha have got some resemblance to the nanoparticle effect.

The metal used for making vessels used for preparing, storing and serving food and water impart some useful properties to the food/water that has been ingested. For example, it has been observed that hot milk drunk from a silver tumbler has some effect on cataract. It now looks probable that nanosized particles from the metal get dissolved in the liquid to cause the effect after ingestion.

Some practices that have been followed from ancient times in Tamil Nadu and are in existence even today include:

1. Manually fanning ourselves using peacock feathers or palmyra leaves.
2. Sleeping on mats made of a particular type of grass.
3. Serving food on plantain leaves.
4. Having an oil bath regularly.
5. Using medicated oils.
6. Smearing sandal paste on the entire body.
7. Use of camphor for pujas and in the preparation of medicines.

All these practices seem to have some nanoparticle effect that has useful effects on the health of humans. Deeper research into these may bring to fore the nanophysiology and nanobiochemistry angle of the ancient practices.

Homeopathic Medicine

Homeopathic medicine (HM) is based on the concept that a material which is supposed to be the cause for disease development, if given in minute quantities can cure the disease. This is similar to the concept of nanomedicine. So, homeopathic medicine may be considered as ancient nanomedicine.

During his career, the author has observed that a wound caused by chemotherapeutic agents at the injected site (given as a cure for carcinoma of the stomach) had restrictive property with respect to progression of cancer. This was presented at the Surgeon's Conference in 1992 at Tirunelveli, Tamil Nadu. The same property was discussed in a book dealing with homeopathic medicine.

Some concepts in nanotechnology seem to be extrapolations of concepts in HM. So, the observations made by the pioneers of HM are nothing but scientific and it needs more research to establish that HM is a form of Nanomedicine, that will create a new and economical treatment for the masses in developing countries to use and benefit from.

APPENDIX V

KIM ERIC DREXLER

Kim Eric Drexler was born on April 25, 1955 in Oakland, California. He is an engineer best known for popularizing the potential of molecular nanotechnology (MNT), from the 1970s and 1980s. Drexler was very strongly influenced by ideas on "Limits to Growth" in the early 1970s. He fabricated metal films, a few tens of nanometres thick on a wax support to demonstrate the potentials of high performance solar sails.

Engines of Creation

The important landmark in the development of molecular nanotechnology was the publication of the book *Engines of Creation* by K. Eric Drexler in 1986. It is a seminal book on molecular nanotechnology.

Engines of Creation is unique for its style and substance. The substance of *Engines of Creation* is extraordinary. Drexler extrapolates a world from the bottom-up where we can build atom by atom. Similarly and inspirationally, physicist Richard Feynman discussed the concept of recursive miniaturization in his 1959 speech "There's Plenty of Room at the Bottom". But only Drexler came up with the idea of using molecular machinery for large-scale fabrication. Drexler sees a world where not only can the entire Library of Congress fit in a chip, the size of a sugar cube, but "universal assemblers" (tiny machines that build atom by atom) will be used for everything from medicinal robots that help clear the capillaries to environmental scrubbers that clear pollutants from the air.

It is in this book that Drexler first published his famous prediction of what might happen if a molecular nanotechnology were used to build uncontrollable self-replicating machines—the "grey goo" scenario.

Molecular Nanotechnology—Origin

During the late 1970s, Drexler began to develop ideas about molecular nanotechnology. In 1979, he encountered Richard Feynman's 1959 talk. The term nanotechnology was coined by the Tokyo Science University Professor Norio Taniguchi in 1974 to describe the precision manufacture of materials with nanometre tolerances and was unknowingly appropriated by Drexler in his 1986 book *"Engines of Creation: The Coming Era of Nanotechnology"* to describe what later became known as molecular nanotechnology. In that book, he proposed the idea of a nanoscale "assembler" which would be able to build a copy of itself and of other items of arbitrary complexity.

His Ph.D. work was the first doctoral degree on the topic of molecular nanotechnology and his thesis, "Molecular Machinery and Manufacturing with Applications to Computation" was published as *"Nanosystems: Molecular Machinery, Manufacturing and Computation"* (1992) and is an important contribution to the field of molecular manufacturing.

Drexler and Christine Peterson founded the Foresight Institute in 1986 with the mission of "Preparing for nanotechnology."

Controversy

Drexler's work on nanotechnology was criticized by Richard Smalley on the following grounds:

1. One of the barriers to achieving molecular nanotechnology is the lack of an efficient way to create machines on a molecular/atomic scale. One of Drexler's early ideas was an "assembler," a nanomachine which would comprise an arm and a computer that could be programmed to build more nanomachines. But the lack of a way to first build an assembler remains the *sine qua non* obstacle for achieving this vision.

2. A second difficulty is design. Hand design of a gear or bearing at the level of atoms is a gruelling task. While Drexler, Merkle and others have created a few designs of simple parts, no comprehensive design effort has been attempted.

3. A third difficulty is separating successful trials from failures and elucidating the failure mechanisms of the failures.

In Darwinian evolution, it proceeds by random variations in ensembles of organisms combined with deterministic reproduction/extinction as a selection process to achieve great complexity after billions of years. Deliberate design and building of nanoscale mechanisms require a means other than reproduction/extinction to winnow successes from failures. Such means are difficult to provide and presently non-existent for anything other than small assemblages of atoms viewable by an AFM or STM.

GLOSSARY

Actuator   A mechanical device for moving or controlling a mechanism or system. It takes energy, usually created by air, electricity, or liquid, and converts it into some kind of motion.

Assembler   A general-purpose device for molecular manufacturing, capable of guiding chemical reactions by positioning molecules.

Atom   The smallest unit of a chemical element, about a third of a nanometre in diameter. Atoms make up molecules and solid objects.

Atomic force microscopy/microscope (AFM) A technique for analysing the surface of a rigid material all the way down to the level of the atom. AFM uses a mechanical probe to magnify surface features up to 100, 000, 000 times, and produces 3D images of the surface. The technique is derived from a related technology, called scanning tunnelling microscopy (STM). The difference is that AFM does not require the sample to conduct electricity, whereas STM does. AFM also works at regular room temperatures, while STM requires special temperature and other conditions. AFM is being used to understand materials problems in many areas including data storage, telecommunications, biomedicine, chemistry, and aerospace. The atomic force microscope was invented in 1986. It uses various forces that occur when two objects are brought within nanometres of each other. An AFM can work either when the probe is in contact with a surface, causing a repulsive force, or when it is a few nanometres away, where the force is attractive.

Atomic layer deposition   A true "nano" technology, allowing ultra-thin films of a few nanometres to be deposited in a precisely controlled way. It is a thin film deposition technique that is based on the sequential use of a gas-phase chemical process. The majority of ALD reactions use two chemicals, typically called precursors. These precursors react with a surface, one at a time, in a sequential manner. By exposing the precursors to the growth surface repeatedly, a thin film is deposited. The two defining characteristics of ALD—self-limiting atomic layer-by-layer growth and highly conformal coating—offer many benefits in semiconductor engineering, MEMS and other nanotechnology applications.

Biocomputing   This can mean at least two different things. First, it can be defined as the construction and use of computers which function like living organisms or contain biological components, and so called biocomputers. In this meaning it is closely related to DNA computing. Biocomputing can also be defined as the use of computers in biological research and in this case it is referred to as Bioinformatics.

Biological assays   Bioassay or biological assay or biological standardization is a type of scientific experiment. Bioassays are essential in the development of new drugs and in monitoring pollutants in the environment. Those are procedures by which the potency or the nature of the substance is estimated by studying its effects on living matter.

Biomedical nanotechnology   The field of nanotechnology that investigates key areas including biomedical applications, understanding of nanomaterials for various biomedical applications, wet and dry nanotechnology, and basics of nanofabrication. It also deals with applications such as drug delivery, imaging and diagnostics, and tissue regeneration and engineering.

Biomimetics   The design of systems, materials, and their functionality to mimic nature, i.e., imitating, copying, or learning from nature. Current examples include layering of materials to achieve the hardness of an abalone shell or understanding why spider silk is stronger than steel.

Bionanotechnology   The integrative technology that aims to use the knowledge gathered from the natural construction of cellular systems for the advancement of science and engineering. Investigating the topology and communication processes of cell parts can lead to invention of novel biological devices with exciting applications.

Biosensor   A device for the detection of an analyte that combines a biological component with a physico-chemical detector component.

Bottom-up methodology   Building organic and inorganic structures atom by atom, or molecule by molecule.

Brownian motion or movement   Motion of a particle in a fluid owing to thermal agitation.

Buckminsterfullerene   A sphere of sixty carbon atoms, also called a buckyball. Named after the architect Buckminster Fuller, who is famous for the geodesic dome that he created and that resembles buckyballs.

Buckyball   A popular name for Buckminsterfullerene.

Catenane   Any compound, especially a hydrocarbon, having two or more rings interconnected like the links of a chain, without a covalent bond.

Cell repair machine   Molecular and nanoscale machines with sensors, nanocomputers and tools, programmed to detect and repair damage to cells and tissues, which could even report back to and receive instructions from a human doctor if needed.

Chemical vapour deposition (CVD)   A technique used to deposit coatings, where chemicals are first vaporized, and then applied using an inert carrier gas such as nitrogen.

Complementary metal-oxide semiconductor (CMOS) The semiconductor technology used in the transistors that are manufactured into most of today's computer microchips.

Composites Combinations of metals, ceramics, polymers, and biological materials that allow multi-functional behaviour. One common practice is reinforcing polymers or ceramics with ceramic fibres to increase strength while retaining light weight and avoiding the brittleness of the monolithic ceramic. Materials used in the body often combine biological and structural functions (e.g. the encapsulation of drugs).

Dendrimer An artificially manufactured or synthesized molecule built up from branched units called monomers. Such processes involve working on the scale of nanometres. Technically, a dendrimer is a polymer, which is a large molecule comprised of many smaller ones linked together.

Diamondoid Structures that resemble diamond in a broad sense; strong stiff structures containing dense, three-dimensional networks of covalent bonds, formed chiefly from first and second row atoms with a valency of three or more. Many of the most useful diamondoid structures will in fact be rich in tetrahedrally coordinated carbon.

Diode A specialized electronic component with two electrodes called the anode and the cathode. Most diodes are made with semiconductor materials such as silicon, germanium, or selenium. Diodes can be used as rectifiers, signal limiters, voltage regulators, switches, signal modulators, signal mixers, signal demodulators, and oscillators.

Dip-pen nanolithography A direct-write soft lithography technique that is used to create nanostructures on a substrate of interest by delivering collections of molecules via capillary transport from an AFM tip to a surface.

Disruptive technology Technology that is significantly cheaper than current, is much higher performing, has greater functionality, and is frequently more convenient to use. This will revolutionize markets by superseding existing technology. "Paradigm shifting" is a well-known connotation. Although the term may sound negative to some, it is in fact neutral. It is only negative when businesses that are unprepared for change fail to adapt, only to fall behind and fail. The results are not evolutionary, they are revolutionary.

DNA Deoxyribonucleic acid. DNA is a code used within cells to form proteins.

DNA chip A purposefully built microchip used to identify mutations or alterations in a gene's DNA.

Dry nanotechnology Derives from surface science and physical chemistry, focuses on fabrication of structures in carbon, silicon, and other inorganic materials. Unlike the 'wet' technology, 'dry' techniques admit use of metals and semiconductors. The active conduction electrons of these materials make them too reactive to operate in a 'wet' environment, but these same electrons provide the physical properties that make 'dry' nanostructures promising as electronic, magnetic, and optical devices. Another objective is to develop 'dry' structures that possess some of the same attributes of the self-assembly that the wet ones exhibit.

Ecophagy   Consuming the biological environment as a result of uncontrolled self-replication. Coined and defined by Robert A. Freitas Jr. Frequently associated with "grey goo," as ecophagy is its main purpose.

Entropy   A measure of the disorder of a closed system. The second law of thermodynamics states that the entropy (and disorder) increases as time moves forward.

Enzymes   Molecular machines found in nature made of protein, which can catalyse (speed up) chemical reactions.

Exponential assembly   A manufacturing architecture starting with a single tiny robotic arm on a surface. This first robotic arm makes a second robotic arm on a facing surface by picking up miniature parts, carefully laid out in advance in exactly the right locations so that the tiny robotic arm can find them and assemble them. The two robotic arms then make two more robotic arms, one on each of the two facing surfaces. These four robotic arms, two on each surface, then make four more robotic arms. This process continues with the number of robotic arms steadily increasing in the pattern 1, 2, 4, 8, 16, 32, 64, etc. until some manufacturing limit is reached. This is an exponential growth rate, hence the name exponential assembly.

Fullerene   A pure carbon molecule composed of at least 60 atoms of carbon. They are cagelike structures of carbon atoms; the most abundant form produced is Buckminsterfullerene (C_{60}), with sixty carbon atoms arranged in a spherical structure. Because a fullerene takes the shape of a soccer ball or a geodesic dome, it is sometimes referred to as a buckyball after the inventor of the geodesic dome, Buckminster Fuller, for whom the fullerene is more formally named.

Genetic algorithm   Any algorithm which seeks to solve a problem by considering numerous possibilities at once, ranking them according to some standard of fitness, and then combining the fittest in some way. In other words, any algorithm which imitates natural selection.

Genomics   The study of the full complement of genes that make up an organism.

Giant magnetoresistance (GMR)   It results from subtle electron-spin effects in ultra-thin 'multilayers' of magnetic materials, which cause huge changes in their electrical resistance when a magnetic field is applied. GMR is 200 times stronger than ordinary magnetoresistance. GMR enables sensing of significantly smaller magnetic fields, which in turn allows hard disk storage capacity to increase by a factor of 20.

Grey Goo   Destructive nanobots. Self-replicating nanomachines, spreading uncontrollably, building copies of themselves using all available material. This is a commonly mentioned nanotechnology disaster scenario, although it is rather unlikely due to energy constraints and elemental abundances. More probable disaster scenarios are the green goo, golden goo, red goo, khaki goo scenarios.

Green Goo Nanomachines or bio-engineered organisms used for population control of humans, either by governments or eco-terrorist groups.

LCD (liquid crystal display) Technology used for displays in notebook and other smaller computers. LCDs allow displays to be much thinner than cathode ray tube technology. LCDs consume much less power because they work on the principle of blocking light rather than emitting it.

LED (light emitting diode) A semiconductor device that emits visible light when an electric current passes through it. The light is not particularly bright, but in most LEDs it is monochromatic, occurring at a single wavelength. The output from an LED can range from red (at a wavelength of $\sim$700 nm) to blue-violet ($\sim$400 nm).

Ligand An ion, a molecule, or a molecular group that binds to another chemical entity to form a larger complex.

Mechanochemistry The coupling of the mechanical and the chemical on a molecular scale and includes mechanical breakage, chemical behaviour of mechanically stressed solids, tribology, polymer degradation under shear, cavitation-related phenomena (e.g. sonochemistry and sonoluminescence), shockwave chemistry and physics, and even the burgeoning field of molecular machines. Mechanochemistry can be seen as an interface between chemistry and mechanical engineering.

Mechanosynthesis Any chemical synthesis in which reaction outcomes are determined by the use of mechanical constraints to direct reactive molecules to specific molecular sites.

Microelectromechanical systems (MEMS) Technology used to integrate various electro-mechanical functions onto integrated circuits. A typical MEMS device combines a sensor and logic to perform a monitoring function. Examples include sensing devices used to control the deployment of airbags in cars and switching devices used in optical telecommunications cables.

Microfluidics The science of designing, manufacturing, and formulating devices and processes that deal with volumes of fluid on the order of nanolitres (symbolized nl and representing units of 10^{-9} litre) or picolitres (symbolized pl and representing units of 10^{-12} litre).

Molecular assembler A general-purpose device for molecular manufacturing, able to guide chemical reactions by positioning individual molecules to atomic accuracy (e.g. mechanosynthesis) and to construct a wide range of useful and stable molecular structures according to precise specifications. Also known as an assembler, a molecular assembler is a molecular machine that can build a molecular structure from its component building blocks.

Molecular beam epitaxy (MBE) Process used to make compound (multi-layer) semiconductors, and consists of depositing alternating layers of materials, layer by layer, one type after another (such as the semiconductors gallium arsenide and aluminium gallium arsenide).

Molecular electronics Any system with atomically precise electronic devices of nanometre dimensions, especially if made of discrete molecular parts rather than the continuous materials found in today's semiconductor devices.

Molecular manufacturing Manufacturing using molecular machinery, giving molecule-by-molecule control of products via positional chemical synthesis, to produce complex molecular structures manufactured to precise specifications.

Molecular medicine A variety of pharmaceutical techniques and gene therapies that address specific molecular diseases or molecular defects in biological systems.

Molecular nanotechnology (MNT) Thorough, inexpensive control of the structure of matter based on molecule-by-molecule control of products and by-products; the products and processes of molecular manufacturing, including molecular machinery; a technology based on the ability to build structures to complex, atomic specifications by mechanosynthesis or other means; most broadly, the engineering of all complex mechanical systems constructed from the molecular level.

Molecular wire A quasi-one-dimensional molecule that can transport charge carriers (electrons or holes) between its ends.

Moore's law The observation made in 1965 by Gordon Moore, co-founder of Intel, that the number of transistors per square inch on integrated circuits had doubled every 18 months since the integrated circuit was invented. Moore predicted that this trend would continue for the foreseeable future.

MWNT Multi-walled nanotubes.

Nano A prefix meaning one billionth (1/1, 000, 000, 000).

Nanoarray An ultra-sensitive, ultra-miniaturized array for biomolecular analysis.

Nanobeads Polymer beads with diameters of between 0.1 to 10 micrometres. Also called nanodots, nanocrystals and quantum beads.

Nanobiotechnology Applies the tools and processes of nano/microfabrication to build devices for studying biosystems in order to learn from biology how to create better nanoscale devices.

Nanocircuits Electrical circuits in the scale of nanometres.

Nanocomposites Polymer/inorganic nanocomposites composed of two or more physically distinct components with one or more average dimensions smaller than 100 nm. From the structural point of view, the role of the inorganic filler, usually as particles or fibres, is to provide intrinsic strength and stiffness while the polymer matrix can adhere to and bind the inorganic component so that forces applied to the composite are transmitted evenly to the filler.

Nanocomputer A computer made from components (mechanical, electronic, or otherwise) built at the nanometre scale.

Nanocrystals Nanoscale semiconductor crystals. Molecular-sized solids formed with a repeating, 3D pattern of atoms or molecules with an equal distance between each part. Nanocrystals are aggregates of anywhere from a few hundred to tens of thousands of atoms that combine into a crystalline form of matter known as a 'cluster'. Typically around 10 nm in diameter, nanocrystals are larger than molecules but smaller than bulk solids and therefore frequently exhibit physical and chemical properties somewhere in between. Nanocrystals are believed to have potential in optical electronics because of their ability to change the wavelength of light.

Nanoelectronics Electronics on a nanometre scale, whether made by current techniques or nanotechnology; includes both molecular electronics and nanoscale devices resembling today's semiconductor devices. Nanotech applications in the field of electronics are especially promising: computer chips, optoelectronics, information storage, nanocomputer, sensors.

Nanofabrication Construction of items using assemblers and stock molecules (nanoscale engineering). Design and manufacture of devices with dimensions measured in nanometres.

Nanofactory A proposed system in which nanomachines (resembling molecular assemblers, or industrial robot arms) would combine reactive molecules via mechanosynthesis to build larger atomically precise parts. These, in turn, would be assembled by positioning mechanisms of assorted sizes to build macroscopic (visible) but still atomically-precise products.

Nanofibres Hollow and solid carbon fibres with lengths in the order of a few microns and widths varying from tens of nanometres to around 200 nm.

Nanofilters Filters useful for the separation of molecules, such as proteins or DNA, for research in genomics. Another use is in water filtration.

Nanofluidics Controlling nanoscale amounts of fluids.

Nanoionics The study and application of phenomena, properties, effects and mechanisms of processes connected with fast ion transport (FIT) in all solid-state nanoscale systems.

Nanolithography Writing on the nanoscale. Derived from the Greek words *Nanos*-"Dwarf", *Lithos* = "rock," and *grapho* = "to write," this word literally means "small writing on rocks." Nanolithography is the art and science of etching, writing, or printing at the microscopic level, where the dimensions of characters are on the order of nanometres. This includes various methods of modifying semiconductor chips at the atomic level for the purpose of fabricating integrated circuits (ICs). Instruments used in nanolithography include the scanning tunnelling microscope (STM) and the atomic force microscope (AFM). Both allow surface viewing in fine detail without necessarily modifying it. Either the STM or the AFM can be used to etch, write, or print on a surface in single-atom dimensions.

Nanomachine An artificial molecular machine of the sort made by molecular manufacturing. They are devices built from individual atoms. Their size is measured in nanometres and they are built from individual atoms. The idea is that the assembler will be able to rearrange atoms from raw material in order to produce useful items.

Nanomanipulation The process of manipulating items at an atomic or molecular scale in order to produce precise structures.

Nanomanufacturing Manufacturing at the nanoscale. It is the industrial application of nanotechnology.

Nanomaterials Materials which have structured components with at least one dimension less than 100 nm. Materials that have one dimension in the nanoscale are layers, such as thin films or surface coatings. Some of the features on computer chips come in this category. Materials that are nanoscale in two dimensions include nanowires and nanotubes. Materials that are nanoscale in three dimensions are particles, for example precipitates, colloids and quantum dots (tiny particles of semiconductor materials). Nanocrystalline materials, made up of nanometre-sized grains, also fall into this category. The focus of nanomaterials is a bottom-up approach to structures and functional effects whereby the building blocks of materials are designed and assembled in controlled ways.

Nanomechanics A branch of nanoscience studying fundamental mechanical (elastic, thermal and kinetic) properties of physical systems at the nanometre scale.

Nanomedicine Application of nanoscience and nanotechnology techniques in the field of medicine. Areas such as disease diagnosis, drug delivery and molecular imaging are being intensively researched. Medical-related products containing nanoparticles are currently produced. The broader categories are 1) the comprehensive monitoring, control, construction, repair, defence, and improvement of all human biological systems, working from the molecular level, using engineered nanodevices and nanostructures; 2) the science and technology of diagnosing, treating, and preventing disease and traumatic injury, of relieving pain, and of preserving and improving human health, using molecular tools and molecular knowledge of the human body; 3) the employment of molecular machine systems to address medical problems, using molecular knowledge to maintain and improve human health at the molecular scale.

Nanometre One nanometre (nm) is equal to one-billionth of a metre, 10^{-9} m, or a millionth of a millimetre. Atoms are below a nanometre in size, whereas many molecules, including some proteins, range from a nanometre upwards.

Nanometrology Nanometrology is the science of measurement at the nanoscale level. Nanometrology has a crucial role in order to produce nanomaterials and devices with a high degree of accuracy and reliability (nanomanufacturing).

Nano-optics Also known as nanophotonics, this refers to interaction of light and matter on the nanoscale.

Nanoparticles Particles less than 100 nm in diameter that exhibit new or enhanced size-dependent properties compared with larger particles of the same material.

Nanopharmaceuticals Nanoscale particles used to modulate drug transport for drug uptake and delivery applications.

Nanoprobe Nanoscale machines used to diagnose, image, report on, and treat disease within the body.

Nanoreplicators A set of nanomachines capable of exponential replication.

Nanorobot A computer-controlled robotic device constructed of nanometre-scale components to molecular precision, usually microscopic in size (often abbreviated as "nanobot"). Also called Nanites.

Nanorods Carbon nanorods; formed from multi-walled carbon nanotubes. Another nanoscale material with unique and promising physical properties, such that they may yield improvements in high-density data storage, and allow for cheaper flexible solar cells.

Nanoscale 1–100 nanometre range. Scale with nanometre order of magnitude.

Nanoscience The study of phenomena and manipulation of materials at atomic, molecular and macromolecular scales, where properties differ significantly from those at a larger scale.

Nanosensor A chemical or physical sensor constructed using nanoscale components, usually microscopic or submicroscopic in size. Nanosensors are any biological, chemical, or sensory points used to convey information about nanoparticles to the macroscopic world. They may be used for various medicinal purposes and as gateways to building other nanoproducts used for computer chips that work at the nanoscale, and nanorobots.

Nanoshells Nanoscale metal spheres that can absorb or scatter light at virtually any wavelength.

Nanostructure Nanostructure is a structure with arrangement of its parts in the nanometre scale.

Nanotechnology Areas of technology where dimensions and tolerances in the range of 0.1 nm to 100 nm play a critical role. It is a manufacturing technology able to inexpensively fabricate most structures consistent with natural law, and to do so with molecular precision. Nanotechnology is the design, characterization, production and application of structures, devices and systems by controlling shape and size at the nanometre scale.

Nanoterrorism Using MNT-derived nanites to do damage to people or places.

Nanotube A one-dimensional fullerene (a convex cage of atoms with only hexagonal and/or pentagonal faces) with a cylindrical shape. Carbon nanotubes discovered in 1991 by Sumio Iijima, are known to act as conductors or semiconductors. Nanotubes are proving to be useful as molecular components for nanotechnology. They are structures which consist of rolled graphene sheets. There are two types of CNT: single-walled (one tube) or multi-walled (more tubes). Both of these are typically a few nanometres in diameter and several micrometres to centimetres long.

Nanowires Semiconductor nanowires are one-dimensional structures, with unique electrical and optical properties, that are used as building blocks in nanoscale devices. Striped or "super latticed" nanowires can function as transistors, LEDs (light-emitting diodes) and other optoelectronic devices, biochemical sensors, heat-pumping thermoelectric devices, or all of the above, along the same length of wire.

NEMS Abbreviation for nanoelectromechanical systems, it is a generic term to describe nanoscale electrical/mechanical devices. Nanoscale MEMS.

Peptide Any of various natural or synthetic compounds containing two or more amino acids linked by the carboxyl group of one amino acid to the amino group of another.

Photolithography The technique used to produce the silicon chips that make up modern-day computers. The traditional process involves shining light through a mask onto a photosensitive polymer (photo resist) on a silicon surface, then subsequently removing the exposed areas.

Photonics Electronics using light (photons) instead of electrons to manage data.

Piezoelectricity The generation of electricity or of electric polarity in dielectric crystals subjected to mechanical stress, or the generation of stress in such crystals subjected to an applied voltage.

Positional assembly A technique that has been suggested as a means to build objects, devices and systems on a molecular scale using automated processes in which the components that carry out the construction process would follow programmed paths. It is a high-precision form of self-assembly.

Productive nanosystems Functional nanometre-scale systems that make atomically specified structures and devices under programmatic control, i.e., they perform manufacturing to atomic precision.

Proteomics Refers to all the proteins expressed by a genome, and thus proteomics involves the identification of proteins in the body and the determination of their role in physiological and pathophysiological functions.

Quantum An indivisible entity of a quantity that has the units as the Planck's constant and is related to both energy and momentum of elementary particles of matter (called fermions) and of photons and other bosons. The word comes from the Latin *quantus*, for "how much." Behind this, one finds the fundamental notion that a physical property may be "quantized", referred to as "quantization". This means that the magnitude can take on only certain discrete numerical values, rather than any value, at least within a range. There is a related term of number.

Quantum chemistry A branch of theoretical chemistry, which applies quantum mechanics and quantum field theory to address issues and problems in chemistry. The description of the electronic behaviour of atoms and molecules as pertaining to their reactivity is one of the applications of quantum chemistry. Quantum chemistry lies on the border between chemistry and physics.

Quantum computer A computer that takes advantage of quantum mechanical properties such as superposition and entanglement resulting from nanoscale, molecular, atomic and subatomic components.

Quantum dot A nanoscale crystalline structure that can transform the colour of light. The quantum dot is considered to have greater flexibility than other fluorescent materials, which makes it suited to use in building nanoscale computing applications where light is used to process information. They are made from a variety of different compounds such as cadmium selenide.

Quantum mechanics A largely computational physical theory explaining the behaviour of quantum phenomena, which incorporates the theory of special relativity.

Quantum tunnelling The passing of electrons through a barrier, without overcoming it or breaking it down.

Quantum well A potential well that confines particles, which were originally free to move in three dimensions, to two dimensions, forcing them to occupy a planar region.

Quantum wire Another form of quantum dot, but unlike the single-dimension "dot", a quantum wire is confined only in two dimensions, that is, it has "length", and allows the electrons to propagate in a "particle-like" fashion. It is constructed typically on a semiconductor base.

Qubit A quantum bit or qubit is a unit of quantum information. It is the quantum analogue of the classical bit.

Rotaxane A mechanically interlocked molecular architecture consisting of a dumbbell- shaped molecule which is threaded through a "macrocycle". The name is derived from the Latin for wheel (*rota*) and axle (*axis*). The two components of a rotaxane are kinetically trapped since the ends of the dumbbell (often called stoppers) are larger than the internal diameter of the ring and prevent disassociation (unthreading) of the components since this would require significant distortion of the covalent bonds.

Scanning electron microscopy (SEM) Utilized in medical science and biology and in such diverse fields as materials development, metallic materials, ceramics, and semiconductors, SEM involves the manipulation of an electron beam that is scanned across the surface of specially prepared specimens to obtain a greatly enlarged, high-resolution image of the specimen's exposed structure. Specimens are scanned with a very fine probe ('tip') and the strength of interaction between the tip and surface is monitored. The specimen can be observed whole for assessing external structure, or freeze-fracture techniques can be used to image internal structures.

Scanning force microscope (SFM)   SFM works by detecting the vertical position of a probe while horizontally scanning the probe or the sample relative to the other. The probe is in physical contact with the sample and its vertical position is detected by detecting the position of a reflected laser beam with a photo diode that consists of two or four segments.

Scanning tunnelling microscope (STM) A device that obtains images of the atoms on the surfaces of materials, and is important for understanding the topographical and electrical properties of materials and the behaviour of microelectronic devices. The STM is not an optical microscope; instead it works by detecting electrical forces with a probe that tapers down to a point only a single atom across. The probe in the STM sweeps across the surface of which an image is to be obtained. The electron shells, or clouds, surrounding the atoms on the surface produce irregularities that are detected by the probe and mapped by a computer into an image. Because of the quantum mechanical effect called "tunnelling", electrons can hop between the tip and the surface. The resolution of the image is in the order of 1 nm or less.

Self-assembly Refers to the use in materials processing or fabrication of the tendency of some materials to organize themselves into ordered arrays (e.g. colloidal suspensions). This provides a means to achieve structured materials "from the bottom-up" as opposed to using manufacturing or fabrication methods such as lithography, which is limited by the measurement and instrumentation

capabilities of the day. For example, organic polymers have been tagged with dye molecules to form arrays with lattice spacing in the visible optical wavelength range and that can be changed through chemical means. This provides a material that fluoresces and changes colour to indicate the presence of chemical species.

Smart materials Reactive materials that combine sensors and actuators, and possibly computers, to enable a response to environmental conditions and changes to those conditions. Examples include uniforms or aircraft skins fabricated from radar-absorbing materials that incorporate avionic links and the ability to modify shape in response to airflow.

Spintronics Electronics that exploits the spin of an electron in some way, rather than just its charge.

Supramolecular chemistry The area of chemistry beyond the molecules, which focuses on the chemical systems made up of a discrete number of assembled molecular subunits or components.

SWNT Single-walled nanotubes.

Tensile strength The maximum amount of tensile stress that can be applied to a material before it ceases to be elastic. If too much force is applied, the material will break or become plastic, i.e., once the force exertion is stopped, the material will not go back to its initial shape.

Tissue engineering The application of the principles and methods of engineering and the life sciences towards the fundamental understanding of structure/function relationships in normal and pathological mammalian tissues and the development of biological substitutes to restore, maintain, or improve functions.

Top-down methodology Making nanoscale structures by machining and etching techniques.

Transistor The basic element in an integrated circuit. An on/off switch (consisting of three layers of a semiconductor material) that consists of a source (where electrons come from), a drain (where they go) and a gate that controls the flow of electrons through a channel that connects the source and the drain. There are two kinds of transistors, the bipolar transistors (also called the junction transistor), and the field effect transistor (FET).

Tribology Study of friction, wear and lubrication of interacting surfaces.

Wet nanotechnology The study of biological systems that exist primarily in a water environment. The functional nanometre-scale structures of interest here are genetic material, membranes, enzymes and other cellular components. The success of this nanotechnology is amply demonstrated by the existence of living organisms whose form, function, and evolution are governed by the interactions of nanometre-scale structures.

Zeolite Any one of a family of hydrous aluminium silicate minerals, whose molecules enclose cations of sodium, potassium, calcium, strontium, or barium, or a corresponding synthetic compound, used chiefly as molecular filters and ion-exchange agents.

Zetta technology The typical number of distinct designed parts in a product made by the systems we envision (molecular, mature, or molecular-manufacturing-based nanotechnology) (zetta means 1021). The term refers to the implemented technology and its products, rather than to intermediate steps on the pathway.

N